BURLEIGH DODDS SCIENCE: INSTANT INSIGHTS

NUMBER 121

# Soil carbon sequestration

burleigh dodds
SCIENCE PUBLISHING

Published by Burleigh Dodds Science Publishing Limited
82 High Street, Sawston, Cambridge CB22 3HJ, UK
www.bdspublishing.com

Burleigh Dodds Science Publishing, 1518 Walnut Street, Suite 900, Philadelphia, PA 19102-3406, USA

First published 2026 by Burleigh Dodds Science Publishing Limited
© Burleigh Dodds Science Publishing, 2026. All rights reserved.

British Library Cataloguing in Publication Data
A catalogue record for this book is available from the British Library

ISBN 978-1-83545-264-6 (Print)
ISBN 978-1-83545-265-3 (ePub)

DOI: 10.19103/9781835452653

Typeset by Deanta Global Publishing Services, Dublin, Ireland

# Contents

# Series list

# Acknowledgements

Chapters in this Instant Insight are taken from the following sources:

Chapter 1 - Introduction: soil carbon sequestration – a process linking soils to humanity
Chapter taken from: Rumpel, C. (ed.), Understanding and fostering soil carbon sequestration, Burleigh Dodds Science Publishing, Cambridge, UK, 2022, (ISBN 978 1 80146 969 5; www.bdspublishing.com)

Chapter 2 - Promoting carbon sequestration in soils: the importance of soil, region and context-specific interventions
Chapter taken from: Rumpel, C. (ed.), Understanding and fostering soil carbon sequestration, Burleigh Dodds Science Publishing, Cambridge, UK, 2022, (ISBN 978 1 80146 969 5; www.bdspublishing.com)

Chapter 3 - Measuring and monitoring soil carbon sequestration
Chapter taken from: Rumpel, C. (ed.), Understanding and fostering soil carbon sequestration, Burleigh Dodds Science Publishing, Cambridge, UK, 2022, (ISBN 978 1 80146 969 5; www.bdspublishing.com)

Chapter 4 - Spectral mapping of soil organic carbon
Chapter taken from: Otten, W. (ed.), Advances in measuring soil health, Burleigh Dodds Science Publishing, Cambridge, UK, 2021, (978 1 78676 426 3; www.bdspublishing.com)

Chapter 5 - Assessing the benefits of temperate agroforestry in enhancing carbon sequestration
Chapter taken from: Mosquera-Losada, M. R., Ladislau, M., Pantera, A. and Chatrchyan, A. (eds.), Advances in temperate agroforestry, Burleigh Dodds Science Publishing, Cambridge, UK, 2025, (ISBN 978 1 80146 719 3; www.bdspublishing.com)

# Chapter 1

## Introduction: soil carbon sequestration – a process linking soils to humanity

*C. Rumpel, CNRS, Sorbonne University, Institute of Ecology and Environmental Sciences Paris, France*

## 1 Introduction

Human survival is threatened by perturbations of the earth system, which are affecting nine key processes, defined as planetary boundaries (Steffen et al., 2015). Most threatening are the perturbations affecting global biogeochemical cycling and, in particular, the carbon cycle. Indeed, fossil fuel combustion and land use change have led to increasing atmospheric $CO_2$ concentrations with severe consequences for the Earth's climate (IPCC, 2022). Climate change is expected to affect production on agricultural soils, which, however, needs to be maintained or even increased in view of a growing global population, because soils deliver as much as 98.8% of our food calories along with multiple other ecosystem services. However, intensification of soil use through modern agriculture is also an important contributor to the perturbations affecting the planetary boundaries (Kopittke et al., 2021).

It has recently been pointed out that, in view of the challenges faced by human societies in particular, soil organic carbon (SOC) plays a central role as a crucial indicator of soil functioning, determining many of the soil's complex and interconnected processes at several scales (Lal et al., 2021; Kopittke et al., 2022). In meeting the objective of providing solutions to global challenges, SOC sequestration has received particular attention. This attention is related to the fact that soils can store huge amounts of carbon and, as part of the global carbon cycle, can influence atmospheric $CO_2$ concentrations responsible for climate change (Fuss et al., 2018). As a consequence, carbon sequestration in soils has been proposed as one of the nature-based solutions to remove $CO_2$ from the atmosphere (Smith, 2016; IPCC, 2022). Indeed, only small increases

http://dx.doi.org/10.19103/AS.2022.0106.01

in total global SOC stocks may be necessary to mitigate fossil fuel emissions (Balesdent and Arrouays, 1999). Although SOC sequestration may be needed to combat climate change, it clearly should not be used as an excuse to not reduce emissions because it cannot be the only climate solution (Chabbi et al., 2017; Anderson et al., 2019). SOC sequestration may be a win-win strategy (Bossio et al., 2020) because increasing SOC is also required to improve all other soil functions with beneficial implications for human societies (Chabbi et al., 2017).

Indeed, SOC sequestration can be seen as underpinning several UN Sustainable Development Goals (SDGs) (Lal et al., 2021; Rumpel et al., 2022) through the provision of food and fiber production, purification of drinking water, biodiversity and climate regulation (FAO et al., 2020; Kopittke et al., 2022). The provision of these ecosystem services relies on the good functioning of soil, in particular its microbial processes responsible for biogeochemical cycling (Lemanceau et al., 2015). In this context, SOC sequestration is an important process, which captures $CO_2$ via photosynthesis and transfers organic materials via plant activity belowground where they can be used by the soil (micro-) organisms. Microbial processes involve the decomposition, transformation and turnover of plant-derived organic materials, providing nutrients, carbon and energy for microbial activity and growth and more generally SOC accumulation. This microbial functioning leads to nutrient provision to plants and aggregate formation, a structure-giving process allowing for water infiltration and storage thereby reducing soil loss through erosion. Moreover, microbial carbon use also leads to microbial necromass formation, a process largely responsible for SOC stabilization (Cotrufo et al., 2013).

## 2 Key issues in understanding soil carbon sequestration

Despite the vital importance of biogeochemical cycling, understanding the processes occurring within the soil matrix is still incomplete. One of the reasons is that soil organic matter is one of nature's most versatile materials comprising a continuum of microbial and plant compounds in various stages of decomposition, which need to be characterized by a combination of methods (Kögel-Knabner, 2002, 2017). In recent years, it has become apparent that a complete understanding of microbial utilization of soil organic compounds is only possible by taking into account the physical organization of the soil matrix determining microbial access (Vogel et al., 2022). The understanding of soil functioning and the processes responsible for ecosystem service provision thus requires the combination of concepts derived from biological, chemical and physical sciences. Information needs to be collected at several scales, from the microscale, where the processes are occurring through to the profile and landscape scales, where soil management takes place (O'Rourke et al., 2015).

Integration and upscaling through different modeling approaches is also needed to understand the importance and consequences of specific processes for biogeochemical cycling at the global scale (Jungkunst et al., 2022). However, global carbon modeling is characterized by a huge uncertainty, making climate predictions at the global scale variable (Jones and Frielingstein, 2020). This is related to the poor understanding of land-based processes to which soil and in particular SOC cycling makes a major contribution. Biogeochemical and biophysical understanding of soil biogeochemical cycling may help to reduce the uncertainty in modeling the global carbon cycle (Bradford et al., 2016).

However, SOC is not the only carbon type in soil. Especially in arid regions and soils with limestone materials, inorganic carbon contributes greatly to soil carbon. It has been estimated that inorganic carbon represents about one-third of the total amount of carbon stored in the soil, amounting to between 700 and 940 GT (Monger et al., 2015). Particularly in regions with poor vegetation growth, where weathering processes are dominant, mineral weathering may be one process which could lead to inorganic carbon sequestration through chemical weathering of Ca-silicate and carbonate rocks (Liu et al., 2011). The cycling of organic and inorganic carbon is affected by fundamentally contrasting processes. While organic carbon is mainly cycled by biological processes relatively rapidly, chemical processes control the slow cycling of inorganic carbon. Interactions between the two carbon pools may occur through biologically enhanced weathering (Song et al., 2018) and pedogenic carbonate formation (Zamanian et al., 2016), two processes, which have poorly been described up to now. Management of soils for organic carbon sequestration may lead to loss of soil inorganic carbon, which therefore needs to be taken into account (Raza et al., 2021).

In assessing global SOC stocks, it must be considered that the up to 3000 GT of carbon stored in global soils (Scharlemann et al., 2014) are not homogenously distributed. More than one-fifth of the world's SOC is stored in organic soils such as peatlands (Leifeld and Menichetti, 2018). These organic soil types are characterized by the occurrence of completely different processes compared to those occurring in mineral soils (Rumpel, 2019). On the other hand, organic soils occupy only about 3% of the total land surface (Yu et al., 2010). Most agricultural production occurs on mineral soil types, which are generally impoverished in SOC and therefore present a huge potential for additional SOC sequestration, especially in deep soil horizons which have been poorly studied (Rumpel and Kögel-Knabner, 2011) but present possibilities of increasing soil carbon with appropriate management interventions (Button et al., 2022).

Understanding the processes affecting the quantity, nature and turnover of organic matter in soils, but also of inorganic carbon and biogeochemical cycling more generally, is thus of crucial importance to find solutions to reducing carbon loss and increasing its sequestration. Specific management practices are needed to influence the different processes affecting contrasting carbon forms in the various soil types (Kögel-Knabner and Amelung, 2021). Moreover, the scope of research on soil carbon has evolved in recent years because soil carbon sequestration relates to land management practices and thus to socioeconomic and political factors. Soil carbon sequestration concerns management decisions affecting biogeochemical processes and is intimately linked to human societies. Its understanding and implementation requires a transdisciplinary approach bringing together different scientific disciplines and various stakeholders (Chabbi et al., 2021; Rumpel et al., 2022). Knowledge gaps are manifold. They range from factors controlling biophysical processes in different pedoclimatic environments to socioeconomic factors influencing the behavior of land managers and political decision-makers at different levels.

To develop adequate policies to enhance good soil management, leading to maintenance and accumulation of soil carbon, the effect of management practices on (bio-)geochemical cycling needs to be taken into account. In view of the numerous soil types and pedoclimatic conditions in the Earth system, the practices to be employed need to be region specific, taking into account not only biophysical constraints and trade-offs but also the socioeconomic, legal and cultural environment (Amelung et al., 2020).

## 3 This book

This book is a first attempt to gather the knowledge of the global (soil) science community working on the different aspects of soil carbon sequestration. It is divided into four parts.

The first part describes the biological, physical and chemical processes determining soil carbon sequestration and their implications for soil management. The second part is devoted to measuring, reporting and verification (MRV) of carbon sequestration in soils. These aspects are crucial for developing management strategies with positive effects on soil carbon. Only if something is measurable can it be influenced by human activity and taken into account by policymakers. However, due to its heterogeneous nature, distribution and residence time, MRV and modeling of soil carbon are quite complex.

In the third part of the book, management practices influencing the different carbon types present in various ecosystems are presented and discussed, while the fourth part of the book is devoted to socioeconomic, legal and policy aspects of soil carbon sequestration. The latter presents global (1) and national policy frameworks for carbon credits related to soil management, (2) the approaches needed to take into account farmers' practices and (3) legal requirements for implementing SOC sequestration.

I would like to thank the authors of the 29 chapters of the book for their valuable contributions. Special acknowledgments go to the reviewers, David Whitehead, Gilles Lemaire, John M. Galbraith and Klaus Lorenz. I hope that this book will be of use to students, scientists and practitioners and will lead to better management of global (soil) carbon resources.

# 4 References

Amelung, W., Bossio, D., de Vries, W., Kögel-Knabner, I., Lehmann, J., Amundson, R., Bol, R., Collins, C., Lal, R., Leifeld, J., Minasny, B., Pan, G., Paustian, K., Rumpel, C., Sanderman, J., van Groenigen, J. W., Mooney, S., van Wesemael, B., Wander, M. and Chabbi, A. 2020. Towards a global-scale soil climate mitigation strategy. *Nature Communications* 11(1), 5427.

Anderson, C. M., Defries, R. S., Utterman, R., Matson, P. A., Nepstad, D. C., Pacala, S., Schlesinger, W. H., Shaw, M. R., Smith, P., Weber, C. and Field, C. B. 2019. Natural climate solutions are not enough. *Science* 363(6430), 933-934.

Balesdent, J. and Arrouays, D. 1999. Usage des terres et stockage de carbone dans les sols du territoire français. Une estimation des flux nets annuels pour la période 1900-1999. C. R. Acad. *Agriculturists* 85, 265-277.

Bossio, D. A., Cook-Patton, S. C., Ellis, P. W., Fargione, J., Sanderman, J., Smith, P., Wood, S., Zomer, R. J., von Unger, M., Emmer, I. M. and Griscom, B. W. 2020. The role of soil carbon in natural climate solutions. *Nature Sustainability* 3(5), 391-398.

Bradford, M., Wieder, W., Bonan, G., Fierer, N., Raymond, P. A. and Crowther, T. W. 2016. Managing uncertainty in soil carbon feedbacks to climate change. *Nature Climate Change* 6, 751-758.

Button, E. S., Pett-Ridge, J., Murphy, D. V., Kuzyakov, Y., Chadwick, D. R. and Jones, D. L. 2022. Deep-C storage: Biological, chemical and physical strategies to enhance carbon stocks in agricultural subsoils. *Soil Biology and Biochemistry* 170, 108697.

Chabbi, A., Kögel-Knabner, I. and Rumpel, C. 2021. Soil Science in transition - Re-defining its role under the 4p1000 initiative. *Geoderma* 385, 114891.

Chabbi, A., Lehmann, J., Ciais, P., Loescher, H. W., Cotrufo, M. F., Don, A., SanClements, M., Schipper, L., Six, J., Smith, P. and Rumpel, C. 2017. Aligning agriculture and climate policy. *Nature Climate Change* 7(5), 307-309.

Cotrufo, M. F., Wallenstein, M. D., Boot, C. M., Denef, K. and Paul, E. 2013. The microbial efficiency-matrix stabilization (MEMS) framework integrates plant litter decomposition with soil organic matter stabilization: Do labile plant inputs form stable soil organic matter? *Global Change Biology* 19(4), 988-995.

FAO, ITPS, GSBI, SCBD, EC. 2020. *State of Knowledge of Soil Biodiversity - Status, Challenges and Potentialities*. Rome: Food and Agriculture Organization.

Fuss, S., Lamb, W. F., Callaghan, M. W., Hilaire, J., Creutzig, F., Amann, T., Beringer, T., de Oliveira Garcia, W., Hartmann, J., Khanna, T., Luderer, G., Nemet, G. F., Rogelj, J., Smith, P., Vicente, J. L. V., Wilcox, J., del Mar Zamora Dominguez, M. and Minx, J. C. 2018. Negative emissions—Part 2: Costs, potentials and side effects. *Environmental Research Letters* 13(6), 063002.

IPCC. 2022. Mitigation of climate change. Contribution of Working Group III to the sixth assessment report of the intergovernmental panel on climate change [P.R. Shukla, J.

Skea, R. Slade, A. Al Khourdajie, R. van Diemen, D. McCollum, M. Pathak, S. Some, P. Vyas, R. Fradera, M. Belkacemi, A. Hasija, G. Lisboa, S. Luz, J. Malley, (eds.)]. Summary for policymakers. In: *Climate Change*. Cambridge, UK and New York, NY: Cambridge University Press.

Jones, C. D. and Friedlingstein, P. 2020. Quantifying process-level uncertainty contributions to TCRE and carbon budgets for meeting Paris Agreement climate targets. *Environmental Research Letters* 15(7), 074019.

Jungkunst, H. F., Goepel, J., Horvath, T., Ott, S. and Brunn, M. 2022. Global soil organic carbon-climate interactions: Why scales matter. *WIREs Climate Change* 13(4). DOI: 10.1002/wcc.780.

Kögel-Knabner, I. 2002. The macromolecuar organic composition of plant and microbial residues as inputs to soil organic matter. *Soil Biology and Biochemistry* 34(2), 139–162.

Kögel-Knabner, I. 2017. The macromolecular organic composition of plant and microbial residues as inputs to soil organic matter: Fourteen years on. *Soil Biology and Biochemistry* 105, A3–A8.

Kögel-Knabner, I. and Amelung, W. 2021. Soil organic matter in major pedogenic soil groups. *Geoderma* 384, 114785.

Kopittke, P. M., Berhe, A. A., Carrillo, Y., Cavagnaro, T. R., Chen, D., Chen, Q., Román Dobarco, M., Dijkstra, F. A., Field, D. J., Grundy, M. J., He, J., Hoyle, F. C., Kögel-Knabner, I., Lam, S. K., Marschner, P., Martinez, C., McBratney, A. B., McDonald-Madden, E., Menzies, N. W., Mosley, L. M., Mueller, C. W., Murphy, D. V., Nielsen, U. N., O'Donnell, A. G., Pendall, E., Pett-Ridge, J., Rumpel, C., Young, I. M. and Minasny, B. 2022. Ensuring planetary survival: The centrality of organic carbon in balancing the multifunctional nature of soils. *Critical Reviews in Environmental Science and Technology*, 1–17. DOI: 10.1080/10643389.2021.2024484.

Kopittke, P. M., Menzies, N. W., Dalal, R. C., McKenna, B. A., Husted, S., Wang, P. and Lombi, E. 2021. The role of soil in defining planetary boundaries and the safe operating space for humanity. *Environment International* 146, 106245.

Lal, R., Bouma, J., Brevik, E., Dawson, L., Field, D. J., Glaser, B., Hatano, R., Hartemink, A. E., Kosaki, T., Lascelles, B., Monger, C., Muggler, C., Ndzana, G. M., Norra, S., Pan, X., Paradelo, R., Reyes-Sánchez, L. B., Sandén, T., Singh, B. R., Spiegel, H., Yanai, J. and Zhang, J. 2021. Soils and sustainable development goals of the United Nations (New York, USA): An IUSS perspective. *Geoderma Regional* 25, e00398.

Leifeld, J. and Menichetti, L. 2018. The underappreciated potential of peatlands in global climate change mitigation strategies. *Nature Communications* 9(1), 1071.

Lemanceau, P., Maron, P.-M., Mazurier, S., Mougel, C., Pivato, B., Plassart, P., Ranjard, L., Revellin, C., Tardy, V. and Wipf, D. 2015. Understanding and managing soil biodiversity: A major challenge in agroecology. *Agronomy for Sustainable Development* 35(1), 67–81.

Liu, Z., Dreybrodt, W. and Liu, H. 2011. Atmospheric $CO_2$ sink: Silicate weathering or carbonate weathering? *Applied Geochemistry* 26 (Suppl.), S292–S294.

Monger, H. C., Kraimer, R. A., Khresat, S., Cole, D. R., Wang, X. and Wang, J. 2015. Sequestration of inorganic carbon in soil and groundwater. *Geology* 43(5), 375–378.

O'Rourke, S. M., Angers, D. A., Holden, N. M. and McBratney, A. B. 2015. Soil organic carbon across scales. *Global Change Biology* 21(10), 3561–3574.

Raza, S., Zamanian, K., Ullah, S., Kuzyakov, Y., Virto, I. and Zhou, J. 2021. Inorganic carbon losses by soil acidification jeopardize global efforts on carbon sequestration and climate change mitigation. *Journal of Cleaner Production* 315, 128036.

Rumpel, C. 2019. Soils linked to climate change. *Nature* 572(7770), 442–443.

Rumpel, C. and Kögel-Knabner, I. 2011. Deep soil organic matter – a key but poorly understood component of terrestrial C cycle. *Plant and Soil*, 338, 143-158.

Rumpel, C., Amiraslai, F., Bossio, D., Chenu, C., Henry, B., Fuentes Espinoza, A., Koutika, L.-K., Ladha, J., Madari, B., Minasni, B., Olaleye, A. O., Shirato, Y., Sall, N. S., Soussana, J.-F. and Varela-Ortega, C. 2022. The role of soil carbon sequestration in enhancing human resilience in tackling global crises including pandemics. *Soil Security*, 8, 100069.

Scharlemann, J. P. W., Tanner, E. V. J., Hiederer, R. and Kapos, V. 2014. Global soil carbon: Understanding and managing the largest terrestrial carbon pool. *Carbon Management* 5(1), 81–91.

Smith, P. 2016. Soil carbon sequestration and biochar as negative emission technologies. *Global Change Biology* 22(3), 1315–1324.

Song, Z., Liu, H., Strömberg, C. A. E., Wang, H., Strang, P. J., Yang, X. and Wu, Y. 2018. Contribution of forests to the carbon sink via biologically-mediated silicate weathering: A case study of China. *Science of the Total Environment* 615, 1-8.

Steffen, W., Richardson, K., Rockström, J., Cornell, S. E., Fetzer, I., Bennett, E. M., Biggs, R., Carpenter, S. R., de Vries, W., de Wit, C. A., Folke, C., Gerten, D., Heinke, J., Mace, G. M., Persson, L. M., Ramanathan, V., Reyers, B. and Sörlin, S. 2015. Sustainability. Planetary boundaries: Guiding human development on a changing planet. *Science* 347(6223), 1259855.

Vogel, H.-J., Balseiro-Romero, M., Kravchenko, A., Otten, W., Pot, V., Schlüter, S., Weller, U. and Baveye, P. C. 2022. A holistic perspective on soil architecture is needed as a key to soil functions. *European Journal of Soil Science* 73(1), e13152.

Yu, Z. C., Loisel, J., Brosseau, D. P., Beilman, D. W. and Hunt, S. J. 2010. Global peatland dynamics since the last glacial maximum. *Geophysical Research Letters* 37(13), L13402.

Zamanian, K., Pustovoytov, K. and Kuzyakov, Y. 2016. Pedogenic carbonates: Forms and formation processes. *Earth-Science Reviews* 157, 1-17.

# Chapter 2

## Promoting carbon sequestration in soils: the importance of soil, region and context-specific interventions

Rattan Lal, CFAES Rattan Lal Center for Carbon Management and Sequestration, The Ohio State University, USA

## 1 Introduction

The global carbon (C) cycle (GCC), critical to the moderation of the Earth's climate, consists of diverse C pools (Fig. 1). Prominent among these are atmospheric C (gray), coastal C (blue), soil organic C or SOC (brown), soil inorganic C or SIC

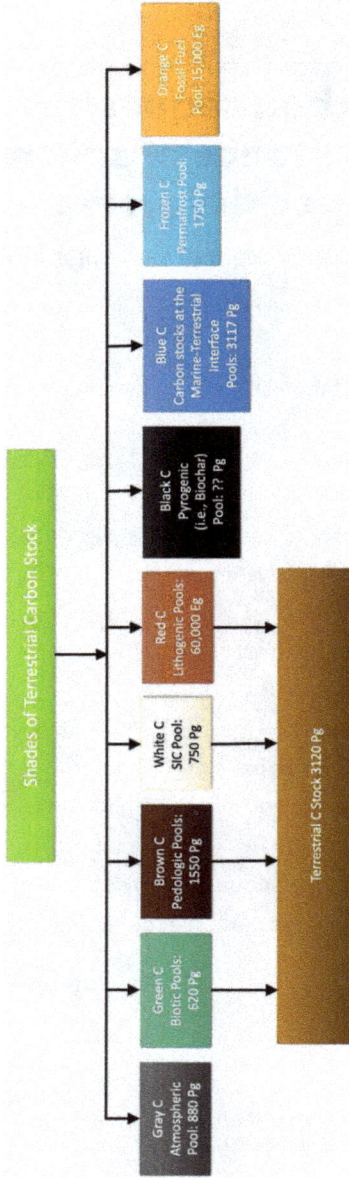

**Figure 1** Different reservoirs of carbon, which interact with each other and are affected by anthropogenic activity. The data on Blue C from Atwood et al. (2020), terrestrial C and black C stock from Lal (2004), gray C from Lal (2018) and frozen C from Jungkunst et al. (2012).

(white), permafrost or Cryosol C (frozen), pyrogenic or fire-derived C (black), vegetation or biomass C (green), lithospheric C (red) and fossil fuel C (orange). Exchange of C among these diverse pools moderates the Earth's climate by influencing the gray or atmospheric C, which in turn strongly impacts the radiation balance of the Earth. Anthropogenic activities, especially deforestation and agriculture along with drainage of peatland and the irrigation of arid land (Aridisols), have drastically perturbed the GCC and set in motion the so-called Anthropocene, which began with the evolution of settled agriculture about 10 millennia ago (Ruddiman, 2003). Anthropogenic influence on the climate was accelerated by the onset of the Industrial Revolution around 1750 and has peaked during the twenty-first century (Crutzen and Stoermer, 2000).

The terrestrial C reservoir (brown, white, green, blue, black and frozen C pools combined) is the third largest pool, after the lithospheric (red) and oceanic reservoirs. In ice-free uplands, the terrestrial C pool comprises soil organic matter (SOM), SIC and vegetation pools. The SOM is composed of ~50% C, and most mineral soils have SOM contents of less than 5%. Yet, SOC has a strong effect on soil quality/functionality and agronomic productivity. The storage and dynamics of SOM, along with its different pools (e.g. labile, intermediate and passive) affect the soil structure and pore size distribution, the aeration and gaseous fluxes between the soil and the atmosphere, water retention and movement, nutrient retention and availability to plants, nutrient and energy availability for soil biota (micro, meso and macro). Indeed, SOM is the bloodstream, essence or elixir of all terrestrial life. Above all, as a major component of the terrestrial C pool, it is an important component of the GCC, which along with the coupled cycling of water ($H_2O$) and other elements (e.g. nitrogen, or N, and phosphorus, or P) forms the basis of numerous ecosystem services (ESs) essential to human well-being and nature conservancy. The decoupling of these critical cycles due to the impact of human activities has created some notable environmental problems including accelerated water runoff and soil erosion by water and wind, eutrophication of water along with the issues of algal bloom and anoxia, contamination of air and the alteration of atmospheric chemistry and changes in the radiative forcing of greenhouse gases (GHGs).

The objective of this chapter is to describe the distribution and dynamics of diverse components of the soil C or pedological reservoir (brown, white, blue, black, green and frozen C), discuss estimates of these pools, explain the impact of anthropogenic activities and outline management strategies to restore them through land use and management options. The literature on soil C reserves is voluminous. Therefore, rather than providing a comprehensive review, the focus of this chapter is on outlining critical features of this urgent topic and on discussing concepts of management that may be used to restore the depleted pedologic C reservoir, with a specific focus on SOC.

**Table 1** Some estimates of SOC and SIC stocks for different ecosystems

| Soil | Location | Country | Depth (cm) | Units | SOC | SIC | Reference |
|---|---|---|---|---|---|---|---|
| Saline | Yellow River Delta | China | 100 | Kg C/m² | 2.3–11.7 | 13.3–24.7 | Zhang et al. (2021) |
| Desert lands | Rajasthan | India | 90 | Mg C/ha | 15.5 | 76.7 | Moharana et al. (2021) |
| Loess Plateau | - | China | 100 | Kg C/m² | - | 17.04 | Tan et al. (2014) |
| Tropical forest succession | Kalimantan | Indonesia | 40 | Mg C/ha | | | |
| | (i) Imperata | | | | 36.2 | | |
| | (ii) Secondary forest | | | | 39.0 | | |
| | (iii) Primary forest | | | | 33.2 | | |
| Blue C | Pacific NW Coast | USA | 300 | Mg C/ha | | | Kauffman et al. (2020) |
| | (i) Seagrass | | | | 217 ± 60 | | |
| | (ii) Low marsh | | | | 417 ± 70 | | |
| | (iii) High marsh | | | | 551 ± 47 | | |
| | Total | | | | 1064 ± 38 | | |
| | Amazon (Mangroves) | Brazil | ? | Mg C/ha | | | Kauffman et al. (2018) |
| | (i) Salt marshes | | | | 257 | | |
| | (ii) Mangroves | | | | 361–746 | | |
| | (iii) AGB | | | | 145 | | |
| | (iv) Sal | | | | 340 | | |
| | Florida Evergreens | USA | | Mg C/ha | | | Reithmaier et al. (2021) |
| | (i) Mangroves | | | | - | 103 ± 61 | |
| | (ii) Marshes | | | | - | 30 ± 30 | |
| | (iii) Seagrass | | | | - | 39 ± 1 | |

| | Location | | | | | Reference |
|---|---|---|---|---|---|---|
| Riverine marginal wetlands | China | 100 | Kg C/m² | 4.6-13.2 | – | Bai et al. (2020) |
| Coniferous forest Taiga | Karelia Isthmus Bryanak Regions | Russia | 50 | Mg C/ha | 47-116 | – | Kuznetsova et al. (2020) |
| Tundra | Arctic | Greenland | | Kg C/m² | 34.9-54.6 | – | Bradley-Cook and Virginia (2018) |
| Histosols | Babia Gora National Park | Poland | ? | Mg C/ha | 50-905 (7550) | – | Reyna-Bowen et al. (2019) |
| Tropical rain forest | Amazon | Brazil | | Mg C/ha | (Total stock 267.5) | – | Dantas et al. (2021) |
| (i) AGB | | | | | 35.2% | |
| (ii) Soil | | | | | 53.2% | |
| (iii) Root | | | | | 1.5% | |
| Grasslands | Pampas | Argentina | | Mg C/ha | 56-59 | – | Alvarez et al. (2021) |

SOC = Soil organic carbon, SIC = Soil inorganic carbon.

## 2 Global distribution of the soil carbon pool among soil types and ecoregions

The magnitude and dynamics of SOC differ widely depending on soil characteristics and region. SOC and SIC stock estimates are highly variable (Table 1), leading to uncertainty in erroneous interpretation of and contradictory conclusions from the data. Examples of the effects of the diverse variables affecting different SOC pools are outlined in this table. Major soil properties affecting SOC are texture (clay and silt content) (Arrouays et al., 2006; Chatterjee, 2020; Djagbletey et al., 2018; Larionova et al., 2008; Okpoho, 2018; Wattel-Koekkoek et al., 2003), mineralogy (Paul et al., 2008; Wattel-Koekkoek et al., 2003; Yu et al., 2013), soil depth (Tan et al., 2014), the soil wetness index (Mishra et al., 2021; Tomar and Baishya, 2020) and climate, comprising of mean annual precipitation (MAP) (Yu et al., 2013; Chatterjee, 2020), mean annual temperature (MAT) (Chen et al., 2013) and other climate variables (Yu et al., 2013). The SOC pool is also affected by interaction among these complex variables, and understanding the primary controls of SOC stocks and their dynamics is critically important.

Chatterjee (2020) measured the mineralizable SOC pool in six ecological regions of the United States and found that the pools varied with climate, vegetation and soil properties. The permanganate oxidizable C (KMNO$_4$-C) ranged from 7.9 mg/g in the state of Florida to 325 mg/g in Hawaii. Chatterjee observed that the MAP was an important factor for SOC and the resistant C pools, while the clay content was related to the mineralizable active C pools and bacterial abundance.

Gruba et al. (2014) assessed the influence of different parent materials on the spatial distribution of SOC stocks in some forest soils. The range of parent materials studied included Tertiary clay stones and sandstones and Quaternary sands of different origins. The <0.05 mm fraction, as a function of the parent material, was an important control for the distribution of C stocks in soils. The SOC content in the mineral fraction increased with an increase in the fines fraction and, in the O horizon, with an increase in the sand content. Furthermore, the SOC stock was also affected by the forest species and the biomass volume (Gruba et al., 2014). All other factors remaining the same, SOM content is strongly related to climate.

Both SOC and SIC stocks vary significantly with depth, as is exemplified by the data in Table 2. Whereas C stocks in subsoil (below 50 cm depth) do not change rapidly (except in soils prone to accelerated erosion and truncation of the topsoil), they must be accounted for to obtain accurate estimates of the C budget. In the Loess Plateau of China, Yu et al. (2020) assessed SOC and SIC stocks to 5 m depth for diverse land use systems (Table 2). Both SOC and SIC stocks were 1.6- and 2.1-fold higher within a 1–5 m depth range as compared

**Table 2** Depth distribution of soil organic and inorganic C for diverse land uses in the Loess Plateau of China

| Land use | 0–5 m depth (Kg C/m²) | |
| --- | --- | --- |
| | SOC | SIC |
| Shrubland | 17.2 | 90.6 |
| Grassland | 16.3 | 104.1 |
| Forest land | 15.2 | 96.2 |
| Cropland | 14.1 | 82.4 |
| Gully land | 6.4 | 50.3 |

SOC = Soil organic carbon, SIC = Soil inorganic carbon.Source: Adapted from Yu et al. (2020).

with depths of 0–1 m, and both stocks decreased significantly from the 5th to the 15th year of cultivation in cropland. Management effects and rather rapid changes of deep C were observed in several studies. However, to date, the processes leading to the accumulation and loss of subsoil C have been poorly understood. Progress is prevented through these diverse and strongly interacting factors and different methodologies, leading to large uncertainties of global SOC stock estimates to 1 m depth and indicating the urgent need for an improved database by introducing a standardized methodology.

# 3 Soil carbon persistence

Persistence, as indicated by the mean residence time (MRT) of soil C stock, is a critical factor in the mitigation of climate change and is determined by a range of interacting biotic and abiotic factors. Indeed, there are numerous uncertainties in the MRT of SOC. Sierra et al. (2018) distinguished between persistence (age) and the transit time (MRT) and argued that C stored in soils is in general much older than that in the release flux, and that there is a large difference between 'mean ages' and the MRTs, with the former being of the order of centuries or millennia and the latter of the order of decades. They concluded that the age distribution may be a better metric to characterize persistence than the MRT (for more information, see Chapter 13 of this book).

In addition to the magnitude of SOC stock, its MRT may also depend on parameters inherent to the soil, for example, its texture. Based on a 40-year-long monoculture of corn grown on a Chernozem, Larionova et al. (2012) found that the mean duration of SOC renewal was 1271–1498 years, but it changed from 697 years in the upper horizon (0–40 cm) to 2742 years in the subsoil (40–60 cm). Furthermore, the MRT increased with decreasing particle size. In addition to particle size or the clay content (Leifeld et al., 2005), the MRT of recently incorporated organic matter (OM) also depends on the type of clay minerals present. Paul et al. (2008) reported that the MRT may be longer in soils dominated by smectite compared with non-crystalline minerals. Paul and colleagues also concluded that the presence of SOC stabilization mechanisms does not necessarily imply that recently incorporated OM is stabilized. Zeraatpishe and Khormali (2013) observed a significant relationship between SOC and the illite and chlorite contents. Wattel-Koekkoek et al. (2003) observed that SOC associated with kaolinite turned over faster (360 years) than that associated with bentonite, and the MRT of the whole clay size fraction can be as large as 1100 years.

There are several mechanisms of protection (e.g. chemical, physical, biological and ecological) (Lal, 2018; Paul et al., 2008; Six et al., 2002, Chapter 2 of this book) of SOC against microbial decay processes. Chemically, some SOC pools are more resistant to decomposition than others. Accordingly, the C:N ratio remains an important determinant of both the persistence and MRT of SOC and the decomposability of particulate OM or POM (Cotrufo et al., 2019). Soil pH is another important control factor for the MRT. For subalpine grasslands, Leifeld et al. (2013) observed that the MRT significantly increased with a decline in pH (5.9 to 3.9). However, differences in MRT along with the pH gradient may not be reflected in SOC stocks because the aboveground biomass (AGB) may be lower under acidic conditions, and relatively higher inputs from AGB may compensate for the faster turnover time at less acidic pH.

Physically, access of SOC to microbial processes may be a more important determinant than its chemical or molecular structure (Dungait et al., 2012). In addition to access, soil temperature is another primary physical (abiotic) determinant of the MRT (Sanderman et al., 2003). Chen et al. (2013) observed a significant declining trend in the global MRT between 1960 and 2008 in soils in high latitude regions characterized by large SOC stocks and a greater degree of climate change. For some French cultivated soils, Balesdent and Recous (1997) observed that 75% of the topsoil C had an MRT of 40 years and coarse fractions contained most of the younger SOC. Further, the MRT may also be related to the MAP (Parshotam et al., 2000).

The source of OM is another important control factor of the MRT. Rasse et al. (2005) argued that understanding the origin of OM stabilization in soils is essential to identifying and promoting management practices that can enhance the accrual of SOC. In this context, the importance of root versus shoot must never be overlooked. Based on the synthesis of the available literature, Rasse et al. observed that the MRT of root-derived C is only partly (~25%) attributable to a higher chemical resistance of the root tissues compared with the shoot tissues. Thus, they proposed that the higher MRT of root-derived C may be due to other factors, such as (a) physio-chemical protection in the deeper horizon, (b) physical protection through mycorrhiza and root hair activities and (c) chemical interactions with metal ions or the formation of so-called organo-mineral complexes. Thus, environmental conditions in the mineral soil, and especially below the A horizon (subsoil), may be an important determinant of the MRT. Krull et al. (2003) emphasized the importance of chemical recalcitrance of black C for SOM protection at very long timescales.

## 4 Temperature sensitivity

The temperature sensitivity of SOC stock and its MRT are major concerns regarding its vulnerability to enhanced decomposition under global warming (see Chapters 5, 8 and 13 of this book). By using the European-wide database, Lugato et al. (2021) estimated an SOC loss of 2.5 ± 1.2 Pg C by 2020 with climate change. Lugato and colleagues recommended coniferous forest management practices to increase plant inputs to soils and offset the vulnerability of the POM fraction. Huang et al. (2018) computed location- and scale-specific correlations between temperature and soil C sequestration at the global scale and observed a negative correlation between temperature and SOC at the regional scale between 52°N and 40°S and a positive one beyond this region. The authors hypothesized that large SOC stocks in low-temperature regions may increase with global warming while the low SOC stocks in high-temperature areas may decrease.

## 5 Soil inorganic carbon

The white C, or SIC, is an important C reservoir not only in arid and semi-arid uplands (Lal, 2019; Table 2 for the Loess Plateau of China) but also in tropical everglades and marshlands. The global magnitude of SIC is estimated at 750 Pg to 1 m and 2300 Pg to 2 m depth (Batjes, 1996, Table 3). In the northern temperate grassland of Canada, Bork et al. (2020) observed that SIC accounted for 34.6% of the total C stock and was particularly large below 30 cm depth. An assessment of a wheat-maize cropland in the North China Plains by Shi et al. (2017) indicated that the SOC content (g/kg) decreased from 5.5 in the 0–20 cm layer to 4.1 in the 20–40 cm layer, but that the SIC content increased from 8.1 in the 0–20 cm layer to 12.1 in the 80–100 cm layer. The SIC stock (g/kg) for horizons deeper than the 100 cm layer was 16.5 ± 4.7 and was more than twice the SOC stock (7.5 ± 1.4). Further, there was a significant positive correlation between SOC and SIC stocks (R = 0.74, P < 0.001) in the cropland of the North China Plains.

Soil management affects SIC in addition to SOC (see Chapter 23 of this book). Zamanian et al. (2021) estimated that, between 1970 and 2020, 0.41 Pg of SIC was released into the atmosphere and 0.72 Pg C will be released as $CO_2$ by 2050. The SIC stock is affected by irrigation, liming and fertilization (see

**Table 3** A wide range of estimates of global soil organic carbon stock to 1 m depth

| Reference | Stock | Range (Pg C ) |
|---|---|---|
| Tifafi et al. (2018) | Total soil C | 2500-3468 |
| Köchy et al. (2015) | SOC | 1230 |
| | SOC (including peatland) | 1325 |
| | SOC tropical soils | 421 |
| | SOC tropical wetlands | 40 |
| | SOC with Histosol $P_b$ value | 1062 |
| Cao and Woodward (1998) | | 1358 |
| Batjes (1996) | | 1550 |
| Eswaran et al. (1993) | | 1505 |
| McNicol et al. (2019) | | 1000 to >3000 |
| Scharlemann et al. (2014) | | 1415 |

$P_b$ = soil bulk density, SOC = soil organic carbon.

Chapter 23 of this book), and the loss of SIC may create a positive, negative or neutral feedback to climate change (Sanderman, 2012), particularly in arid and semi-arid regions.

Similar to the loss of SOC with climate change, the decline of SIC stock may also occur with global warming. For example, based on a study across grasslands in China, Yang et al. (2012) reported an SIC stock decrease between 1980 and 2005 at a mean rate of 26.8 g/m$^2$ · yr in the top 10 cm layer. The dynamic nature of the SIC stock has also been reported in Chernozems (Mikhailova and Post, 2006) in which SIC is a large component of the total C stock below 1 m depth.

# 6 Blue carbon in coastal ecosystems

The global coastline is about 440000 km long and consists of coastal ecosystems that are highly diverse and include intertidal and subtidal areas on and above the continental shelf to a depth of 200 m and on immediately adjacent lands (Burke et al., 2001; Ouillon, 2018). Coastal ecosystems consist of coral reefs, mangroves, tidal wetlands, seagrass beds, barrier islands, estuaries, peatland swamps and so on (Burke et al., 2001). Blue C refers to the C stocks at the marine-terrestrial interface, which includes the variety of ecosystems listed above. The blue C stock includes C in AGB, BGB and sediment C. Atwood et al. (2020) estimated the global mean sedimentary C stocks at 3117 Pg C (3006–3209 PgC) in the top 1 m layer. The authors argued that marine sediments contain a large C reservoir (Table 4) that is vulnerable to climate change. Kauffman et al. (2020) evaluated the total blue C stock of the Pacific Northwest

**Table 4** Examples of some estimates of the global Blue carbon stock

| Reference | Units | Ecosystem carbon stock | Ecosystem |
|---|---|---|---|
| Jardine and Siikamäki (2014) | Mg C/ha | 272 ± 49–703 ± 38 | Mangroves |
| Atwood et al. (2020) | Pg C to 1 m depth | 2239–2391 | Marine sediment |

Coast of the United States at 1064 ± 38 Mg C/ha, accounting for >98% of the total ecosystem C stock (TECS) in seagrass and marsh communities (low elevation, high salinity) and 78% in tidal forests (high elevation and low salinity). Kaufman and colleagues observed that measuring TECS to 100 cm depth, as is usually done, strongly underestimates both C stock and the emission of GHGs from tidal wetlands due to the impact of humans on the climate. In addition to containing large C stocks, coastal wetlands and their blue C provide numerous ESs, highlighting the need for the protection, conservation and restoration of blue C in diverse coastal ecosystems.

Stocks of 'Blue C' vary among different coastal ecosystems (Table 5). In the world's largest mangroves, located in Brazil's Amazon, Kauffman et al. (2018) observed that the mean TECS of the salt marshes was 257 Mg/ha and that of the mangroves ranged from 361 Mg C/ha to 746 Mg C/ha. The C stocks in the Amazon mangroves were over twice those of upland evergreen forests and almost ten-fold those of tropical dry forests (Kauffman et al., 2018). In China, Yang et al. (2020) observed that coastal salt marshes form important reservoirs for SOC, which reservoirs are highly heterogeneous and also system-specific. Seagrass ecosystems are considered among the most efficient C sinks globally. In this context, Ricart et al. (2015) observed that C stocks were ~20% higher inside seagrass patches than at seagrass–sand edges in a *Zostera muelleri* patchy seagrass landscape. Mangrove deforestation can lead to decline in ecosystem C stocks (Ouyang and Lee, 2020). Fu et al. (2021) estimated that

**Table 5** Relative magnitude of SOC and biomass C in different coastal ecosystems

| Ecosystem | C stock (Mg C/ha) | |
|---|---|---|
| | Soil organic carbon | Living biomass carbon |
| Seagrasses | 137.1 | 3.0 |
| Salt marsh | 250.1 | 8.5 |
| Estuarine mangroves | 289.1 | 126.6 |
| Oceanic mangroves | 485.3 | 126.6 |
| Tropical forest | 54.3 | 162.9 |

*Source*: Herr and Landis (2016); IPCC (2013); IUCN (2021); Macreadie et al. (2019); The Blue Carbon Initiative (2019).

Chinese vegetated coastal habitat is a large C sink and stores C in the top 1 m of sediments at an annual rate (Gg C/yr) 44 ± 17 in mangroves, 159 ± 57 in salt marshes and 6 ± 45 in seagrass. Mangrove sediments are a large and critical C reservoir containing 2322 Pg C in the top 1 m (Atwood et al., 2020). Thus, 'Blue C' stored in coastal ecosystems can play an important role in climate change mitigation strategies and thus needs to be protected, conserved and restored (Nehren and Wicaksono, 2018).

## 7 Black or pyrogenic carbon

In this chapter, the term black carbon, or BC (Fig. 1), refers to any fire-derived (pyrogenic) C, including biochar. The latter is derived from the incomplete burning of biomass and has a long MRT (Nguyen et al., 2014). Fire, as a management tool, is widely used in tropical ecosystems, and it has strong pedologic, ecologic, economic and social impacts. Uncontrolled and indiscriminate use of fire can adversely impact the SOC stock and its dynamics, with implications for nutrient loss, volatilization, soil loss by accelerated erosion, loss of biodiversity and a shift in plant communities. A study on the effects of fire at the Bhoramdeo Wildlife Sanctuary of Chhattisgarh, India, by Jhariya and Singh (2021) indicated that SOM and N stocks, micronutrients and microbial biomass C (MBC) were higher in the 0–20 cm layer of no-fire zones than in those under fire treatment. Total soil C stock in the 0–20 cm layer was 69.5 Mg/ha under control compared with 66.5 Mg/ha under medium and 53.7 Mg/ha under low-fire intensity. The total N stock, equal to 4.1 Mg/ha, was also the highest under no-fire treatment (Jhariya and Singh, 2021). However, the MRT of biochar may be overestimated (Lutfalla et al., 2017) because the turnover of biochar may be affected both by properties of the biochar and those of the soil. Similar to pyrolysis for biochar production, the process of hydrothermal carbonization is used to convert biomass into hydrochar, which also leads to C sequestration after soil application. A microcosm study conducted by Gajić et al. (2012) documented that the application of hydrochar as a soil conditioner under field conditions has a moderate potential for SOC sequestration.

Burning of rice straw is practiced widely now in the Indo-Gangetic Plains. This practice has adverse effects on soil and the environment (Shyamsundar et al., 2019). Fire as a management tool was used for millennia in the Yangtze River Delta, China. Lehndorff et al. (2014) examined the contribution of BC derived from burning of rice straw by studying chronosequences of 0–2000 years of rice–wheat rotation compared with adjacent non-paddy systems (50-70 years) in the Bay of Hanzhou, China. The authors observed that, despite its regular long-term input, BC only accounted for 7-11% of total SOC in the surface soil and reached a steady state level in 297 years of 13 Mg of BC/ha, with an MRT of 303 years. They concluded that this low contribution of BC to the total SOC in paddy soils was similar to other global aerobic ecosystems. The fate of C originating from BC input thus remains an open question. Mechanisms leading to its removal from the site of deposition may include fragmentation and loss due to wind and/or water erosion (Rumpel et al., 2015). Effects of fire on SOC stocks under pine forests in the west of Russia (Karelia, Isthamus and Bryonsk regions) were reported to 50 cm depth by Kuznetsova et al. (2020), and it was observed that fire has an important effect on SOC stocks in addition to climate, parent materials, vegetation and agricultural activities.

## 8 Frozen carbon and permafrost soils

Permafrost soils (Cryosols or the frozen C) are a large reservoir of SOC stocks estimated at ~1750 Pg to 3 m depth (Jungkunst et al., 2012). In addition to their vulnerability to climate change and increase in $Q_{10}$ factors (Bradley-Cook and Virginia, 2018; Huntzinger et al., 2020), SOC stocks in permafrost are also limited by a deficiency of nutrients such as P. Based on a study of the response of high Arctic dwarf shrub tundra to 20 years of low-level nutrient regimes, Street et al. (2018) reported that C stocks in vegetation and soil were halved over this period. In comparison, sites where P and N were added, C storage increased by more than 50%. Whereas the addition stimulated decomposition and decreased the C stock, P and N together co-stimulated moss productivity and increased OM accumulation. Street and colleagues hypothesized that if climate warming were to increase P availability, any increase in N enrichment from warming of the Artic may lead to increased C sequestration. If P is limited in the tundra region, increase in N would increase C loss. Temperature-induced melting of tundra soil, studied by Bradley-Cook and Virginia (2018), indicated that frozen soils are a 'hotspot' for soil C storage and $CO_2$ efflux. By analyzing datasets of >2700 profiles, Mishra et al. (2021) reported that the soil wetness index and the elevation are the dominant topographic control parameters (see Table 1), and the surface–air temperature and precipitation are important climate-related determinants of SOC stocks in permafrost soils. Huntzinger et al. (2020) assessed ecosystem C stocks in the Arctic/Boreal region of North America and indicated that errors exist in the secondary data generated by models.

In addition to permafrost, global organic (peat) soils are also an important reservoir of SOC and must be protected against land use change (e.g. drainage, deforestation or clearing, cultivation, fire) and mining. In Switzerland, Leifeld et al. (2005) observed that organic soils accounted for less than 3% of the total area but stored about 28% of the national SOC stock. Drainage and cultivation of organic soils, especially for the development of palm oil plantations in Southeast Asia, have created a large terrestrial C debt and have been a major source of GHG emissions. Moreover, climate change and increased fire frequency in boreal forests may lead to a loss of legacy C in organic soils and thus turn them from a carbon sink into a carbon source (Walker et al., 2019). Global loss of wetlands (Mitsch and Gosselink, 2015), and with them the SOC stored under anaerobic conditions, has been a major ecological issue since the 1950s. Thus, protection of existing wetlands and the rewetting of drained wetlands are high priorities. In China, Bai et al. (2020) evaluated changes in SOC stock in river marginal wetlands and reported that higher flooding frequencies could contribute to an accumulation of soil C and N due to favorable hydrological conditions. Thus, Bai and colleagues concluded that attention must be given to C and N stocks in deeper soils.

## 9 Land use and management

The magnitude of SOC stocks and their dynamics are governed by land use and management. Some researchers have argued that land use may have a greater impact on SOC stocks in temperate cultivated soils than that of climate change over the next 50 years (Balesdent and Recous, 1997). Land use may be an important factor in transforming the labile fraction into passive reservoir with a longer MRT (e.g. perennial land use). In addition to the effect of current land use, historic land use may also affect the SOC stock. However, a study of the impacts of historic land use and forest management over the past 200 years in Europe on SOC stocks by Wäldchen et al. (2013) indicated that there is no consistent trend in relation to the historical management of such lands.

A study of SOC stock in the Amazon Forest by Trumbore and de Camargo (2009) showed that mineral soils (0–30 cm) in the Amazon Basin account for 40 Pg C or 3% of the global SOC stock. When combined with the detrital C and SOC stocks to 1 m depth, the total C stock may be 4 times higher; however, in some cases, it may be extremely vulnerable to future land use and climate change. In Switzerland, Leifeld et al. (2005) reported that about 16% of the national SOC stock has been lost due to cultivation, urbanization and deforestation, and future changes in land use and management may not be sufficient to compensate for this historic loss. A global study by Huang et al. (2018) compared the SOC content of eroded, cultivated, forest and grassland soils. The lowest SOC stock of 55.6 Mg/ha was observed in eroded landscapes, and the highest SOC stock of 83.5 Mg/ha was under forest. To 1.05 m depth, forest soil contained 22.4 Mg C/ha more than cultivated soils. Based on a study on smallholder farms in Abuja, Nigeria, Okpoho (2018) showed that SOC stocks in the 0–20 cm layer amounted to 24.8 ± 0.2 Mg C/ha under cashew, 22.0 ± 0.3 Mg C/ha in maize-cowpea and 13.9 ± 0.1 Mg C/ha under maize monocultures. Given similar baselines at these sites, these results might indicate that the cultivation of cereals with extractive practices can deplete SOC stocks.

## 10 Landscape management

Limiting atmospheric $CO_2$ is a high global priority in the twenty-first century, and sustainable management of the landscape and afforestation are among the preferred options to achieve this. Management of the landscape, comprising both surface features and water resources, is based on a holistic concept. Garten and Ashwood (2002) observed that the effects of topography on soil quality indicators (SOC and SOM stocks and the C:N ratio) were secondary to those of the vegetation/land cover. The indicators were more favorable (e.g. higher stocks and lower ratio) with more land cover. A study conducted on the mountain soils of the Babia Gora National Park, Poland, by Reyna Bowen et al. (2019) showed that topographical factors influenced soil conditions and vegetation cover, which in turn affected C accumulation. In this landscape study, the highest SOC stock was recorded in Histosols situated at landscape positions that allow for high water levels. The SIC density in relation to land cover in the Loess Plateau of China was estimated in the order of grassland > forest land for all soil layers to 1 m depth. Tan et al. (2014) also observed that the highest SIC density at 100 cm depth was observed in alkaline soil and the lowest in subalpine meadow soil. Riparian flood plains, which receive sediments on a regular basis, have a high C sink capacity, for both SOC and SIC, depending on the specific site. Based on a study of a riparian zone in the Yellow River Basin, China, Hou et al. (2021) observed that SOC and SIC stocks to 1 m depth increased at average rates of 2.73 Mg C/ha and 5.54 Mg C/ha · yr and that the SOC and SIC stocks were closely correlated with one another and with the silt content. Jones et al. (2019) hypothesized that stabilizing the atmospheric $CO_2$ concentration can be facilitated by promoting secondary tropical regrowth on abandoned agricultural landscapes. The authors argued that the soil C of abandoned cropland may recover within 40 years of secondary forest growth, but the AGB-C stocks may continue to increase over 100 years.

## 11 Salinity management for enhancing soil organic carbon stocks

Soil salinity is a major issue in semi-arid and arid regions, including irrigated lands (Chapter 23 of this book). Saline soils cover 3.1% of the global land area, or approximately 397 M ha (Setia et al., 2013). Soil salinity affects SOC stock by limiting productivity through adverse changes in pedological processes. Setia and colleagues estimated that, globally, saline soils may have lost an average of 3.4 Mg SOC/ha since they became saline, and they predicted that the world's soils may further lose 6.8 Pg C due to salinity by 2100. Regardless of the salinity, projected global warming may also accelerate the decomposition of SOC stocks. The SOC stock in salinized soil is adversely affected by reduced biomass productivity and changes in biological processes. A study on cropland in the Yellow River Delta, China, by Zhang et al. (2021) showed a large range (kg $C/m^2$) of SOC (2.3-11.7) and SIC (13.3-24.7) over 1 m depth. Zhang and colleagues observed that SOC and SIC stocks were the lowest in alkaline soils with the highest soil pH (8.6-9.0). Further, SOC desorption was greater in saline soils with an alkaline pH, indicating low SOC stability in these soils. These results were corroborated by a study on the alkali-sodic soils of Northeast China, where the SOC stock was negatively correlated with soil pH and exchangeable sodium percentage (ESP) and positively with SIC content (Wang et al., 2020). The authors also observed a significant loss in total C stock in corn fields, although the stock was preserved at 120 Mg C/ha in paddies. They concluded that corn and rice cultivation reduced alkali-sodic conditions and favored SOC accumulation but reduced SIC stocks. Flooding of rice paddies facilitated the leaching of salts and transport of dissolved organic carbon (DOC) into the subsoil.

The DOC, being an easily decomposable, labile fraction, may be leached into the subsoil with seepage or percolating water. Sanderman and Amundson (2008, 2009) observed that DOC leaching is an important process of transporting labile C from surface layers and stabilizing it within the mineral soil in deeper layers, especially in soil types developed under high rainfall conditions. They reported that DOC transported into mineral soil accounts for 22% of the annual C inputs below 40 cm in coniferous forest compared with only 2% of C below 20 cm in prairie soil. The DOC transported may be subsequently absorbed on mineral soil and may have long MRTs of 90–150 years. Based on a study of agroforestry systems in Canada, Lim et al. (2018) observed that DOC concentrations under a silvopasture system were 22–24% higher than in other systems and that the same was also true of the SOC stock. Furthermore, the SOC stock in a 0–30 cm layer of mineral soil was more responsive to changes in land use type than to topography (Lim et al., 2018).

## 12 Conservation agriculture, cover cropping and agroforestry

The projected level of global warming may exacerbate the decomposition of SOC stocks and create a positive feedback to climate change. Jones et al. (2005) estimated that soil C stocks may decrease by 54 Pg C to 80 Pg C by 2100. Thus, there has been an increasing emphasis on the need to sequester atmospheric $CO_2$ into the SOM pool. Strategies to bring about this transformation include those that increase the input of biomass C into soil while reducing the risks of decomposition and erosion (Chapters 16, 20 and 22 of this book). Conservation agriculture, or the no-till system with cover cropping, is one such option (Machado, 2005). All other factors remaining the same, the conversion of plow tillage to a system-based conservation agriculture on a highly erodible landscape can enhance SOC stocks (Lal, 2015). This is especially true in tropical environments (Chapter 22 of this book). Based on a study in the Brazilian Cerrado, de Carvalho et al. (2014) observed that the use of a no-till system cover crop maintains or increases SOC and N stocks and enhances soil fertility. It also results in high concentrations of exchange bases, higher CEC and high base saturation in the surface layer. SOC stocks to 40 cm depth differed among cover crop species and the use of *Mucuna puriens* and *Canavalia brasiliensis* increased SOC and N stocks. *Imperata cylindrica* is a rhizomatic weed that occupies a large area of deforested land in Indonesia and West Africa. van der Kamp et al. (2009) observed that SOC storage in the A horizon increased by 14% from 14.5 g/kg under *Imperata* to 16.5 g/kg under secondary forest. *Imperata* can be controlled by growing an aggressive cover crop (e.g. *Mucuna utilis*) for several years followed by conservation agriculture. The latter, along with the use of cover cropping or pastured forages, can be used in conjunction with agroforestry to enhance SOC stocks. This can enhance the ecosystem C budget even in an arid climate. For example, a study in Rajasthan, India, by Tanwar et al. (2019) showed that the ecosystem C stock (Mg C/ha of biomass and soil) over a period of nine years was in the order of farm forestry (47.6) followed by *Ziziphus*-based systems (32.5–33.9). About 50–78% of the additional soil C stock was in the form of SIC. Tanwar and colleagues observed that total ecosystem C sequestered (Mg C/ ha biomass and soil) decreased in the order farm forestry (49.8) > silvo-arable systems (11.0-13.3) > hortipasture systems (8.3) > agri-horti (5.5), silvopasture (5.4), permanent pasture (5.3) and permanent cropping (1.0).

Site-specific adoption of a system-based conservation agriculture may also offset possible decline in the net primary productivity (NPP) due to projected climate change. In addition to the loss of global SOC stock by 54–80 Pg C because of the decomposition of SOM under a changing climate (Jones et al., 2005), Cao and Woodward (1998) predicted decline in NEP with climate change. Cao and Woodward estimated that global NPP is 57 Pg C/yr, that C stocks in vegetation and soil are 640 Pg and 1358 Pg, respectively, and that the NEP ranges from −0.5 Pg C in October to 1.6 Pg C in July. Cao and Woodward further predicted that, with the doubling of $CO_2$, NPP will increase to 69.6 Pg C/yr, the C stocks in vegetation and the world's soil may increase by 133 Pg (773 Pg) and 160 Pg (1518 Pg), respectively, and the seasonal amplitude in NPP will increase by 71%. Increase in NPP by 25% due to the $CO_2$ fertilization effect and the return of biomass C to soil under a conservation agriculture may accentuate the rate of SOC sequestration with a strong increase in global SOC stocks. Therefore, a global increase in land area under conservation agriculture from 180 M ha at present (Kassam et al., 2019) to a larger area may further enhance the SOC stocks in the world's cropland soils (Lal, 2015). Moreover, a less intensive system under conservation agriculture may also be better adapted to climate extremes and pests, which will most likely increase with climate change.

## 13 Grazing management

Livestock management and grazing are practiced on ~3.5 billion ha globally. Grazing management and the associated practices have strong impacts on soil C stocks, with implications for managing these ecosystems to optimize SOC storage (Smith et al., 2014, Chapter 18 of this book). Based on integrated crop/livestock systems (ICLS) in South Dakota, USA, Polat et al. (2020) observed that ICLS can improve SOC stock and soil quality. Their data showed that a low stocking rate in conjunction with ICLS enhanced SOC stock from 20.7 g/kg to 28.3 g/kg and total N from 2.06 g/kg to 2.60 g/kg for 0-5 cm depth. Similarly, Smith et al. (2014) studied the impact of 3 sheep stocking treatments (2.7 ewe/ha · yr, 0.9 ewe/ha · yr and no grazing) in the Scottish uplands. The SOC stock in *Molinia caerulea* swards were 3.83 Mg C/ha, 5.01 Mg C/ha and 6.85 Mg C/ha under high, low and no stocking, respectively. The authors concluded that no sheep and low-intensity sheep grazing are preferred upland management practices for enhancing soil and plant C stocks than commercial sheep grazing. In contrast to the general belief that grasslands are always a C sink, a study in the Pampas (Argentina) by Alvarez et al. (2021) showed that SOC stocks to 50 cm depth declined from 59 Mg C/ha in 2007 to 56.2 Mg C/ha in 2019. In northern temperate grasslands, Bork et al. (2020) observed that SOC stock increased from 24.7 Mg/ha to 57.4 Mg/ha to 0.6 m depth with an increase in the normalized stocking rate from 0.49 to 2.3 times the recommended rate because of the increased abundance of introduced vegetation, such as the rhizomatous grass *Poa pratensis*. Furthermore, SIC stock did not vary with the stocking rate, but SOC was associated with the long-term stocking rate. Grassland/savannas are widespread ecosystems in Sub-Saharan Africa. Based on a study in the Guinea Savanna region of Ghana, Djagbletey et al. (2018) found that SOC stocks at 0-10 cm depth (4.8-12.6 Mg C/ha) were associated with higher amounts with the silt and clay fraction than with micro- and small macro-aggregates.

## 14 Nutrient management for sequestration of soil organic carbon

Biomass, especially the residues of cereals, have larger C:N, C:P and C:S ratios than the SOC fractions (labile, intermediate and passive). Therefore, conversion of biomass into SOC pool, especially a stable pool with a larger MRT, requires inputs of additional nutrients such as N, P and S (Himes, 1998) and micronutrients. In general, a deficiency of N and P may limit the transformation of biomass C into SOC. Spohn (2020) observed that the storage of SOC in mineral soils leads to sequestration of large amounts of organic P. Spohn hypothesized that the formation of mineral-associated SOM is favorable for storing SOC in soil over decadal to millennial timescales and sequesters a large amount of organic P. For example, storage of 1 Mg of SOC in the clay fraction of the topsoil in croplands sequesters 13.1 kg P/ha (Spohn, 2020). P input is also important to enhancing SOC stock in tundra region permafrost soils (Street et al., 2018). Crowther et al. (2019) studied the sensitivity of global soil C stocks to combined nutrient enrichment and concluded that nitrogen and P enrichment in isolation had a minimal impact on soil C storage. When combined with potassium (K) and micronutrients, however, soil C stock increased on average by 0.04 kg C /m² · yr. Several studies have observed positive correlations of SOC stocks with the C:N ratio of the biomass (Devi and Sherpa, 2019). Jones et al. (2019) observed that soil C stocks in secondary tropical forest growth on abandoned cropland may be positively related to soil N reserves, and they concluded that soil C recovered within 40 years of forest rejuvenation. The studies show the important nutrient requirements for SOC storage in mineral soils. In contrast, the need for P and other nutrients for restoring SOC stocks in the organic soils of peatlands would be much less.

There is a growing trend in the adoption of organic farming and agriculture systems with reduced agrochemical input. The concept is based on restoring SOM content through recycling of biomass, using organic amendments and strengthening the mechanisms that couple the cycling of $H_2O$ with those of C and essential elements (e.g. N, P, S). Leite et al. (2003) compared soil quality after 16 years of using organic vs. mineral fertilizers. While soils under organic farming systems had higher SOC and SON contents for the same soil type compared with unfertilized ones, they also contained more light (POM) or labile fractions. In general, management-induced changes in the POM and labile fractions are more evident than those in total SOC stocks. Although the labile and stable fractions both ensure different soil functions, their optimal proportions for soil function are as yet unknown.

## 15 Carbon sequestration under urban ecosystems

Urbanization is a phenomenon of the twenty-first century, and the rapid rate of urban encroachment is strongly affecting the landscape and pedological processes. However, the historic and current effects of urbanization on SOC stocks and dynamics are not well understood with regard to features such as surface sealing, functional zoning, green spaces, history and so on. Vasenev et al. (2013) evaluated the effects of urbanization in the Moscow (Russia) region on SOC for different land use types, soils, urban zones and the so-called 'cultural layers', the latter being the result of human residential activity and settlement history. Vasenev and colleagues observed that the

SOC stock was highly heterogeneous and differed with soil type, soil depth and land use factor and that the SOC content in urban soil has a 'unique character' that deserves an in-depth study. In a subsequent study, Vasenev and Kuzyakov (2018) reported that urban soils and cultural layers may sequester C for centuries, leading to large C stocks beneath cities. Data from a wide range of ecoregions showed that the total C content in urban soils was 1.5–3.0 times higher and the C accrual was much deeper compared with natural soils, resulting in 3 to 5 times larger C stocks. Furthermore, urban SOC stocks increase with latitude but SIC stocks are less affected by climate. Vasenev and colleagues hypothesized that with the long-term input of C from outside cities, urban soils are global hotspots of long-term C sequestration (see also Chapter 24 of this book).

## 16 Conclusion

Soil, or the pedological carbon (C) reservoir, comprising of the SOC and SIC pools, plays a critical role in the GCC and in moderating the Earth's climate. Both need to be considered when applying soil management practices intended to foster soil carbon. The SOC pool is a strong determinant of soil functionality, especially soil health, through which it affects quality and quantity of agronomic productivity (food and nutritional security) and generates numerous ESs essential to humans and nature. Global hotspots of SOC stocks, in terms of the magnitude and their vulnerability to climate change and anthropogenic perturbations, are in coastal ecosystems, permafrost soils, wetland and peat soils and tropical rainforests. Research priorities should include a strengthening of the database of carbon stocks differentiated by carbon type and their vulnerability in different global regions. This will allow for the identification of areas requiring particular attention and specific management strategies in terms of protecting existing stocks or increasing soil carbon when contents are low.

The literature review supports the following conclusions:

1. Terrestrial and soil C stocks are important components of the GCC and critical to moderating the Earth's climate.
2. Anthropogenic land use and soil/crop/livestock management have depleted soil C stock and caused emission of GHGs into the atmosphere. Agriculture contributes about 25% of global GHG emissions. Thus, most soils of agro-ecosystems are depleted of their SOC stocks.
3. Soil C stocks comprise of SOC and SIC components, and both are important to moderation of the climate. Land use and soil management affect both SOC and SIC stocks.

4   Estimates of the global C stocks are highly variable, obtained by diverse and unstandardized methods and often reported to 30 cm depth. Soil C stocks must be reported to at least 1 m depth and preferably to 2 m and 3 m depths through a standardized methodology. The development of simple tools, which can be used in the field to measure soil health and relate it to soil C stocks and other parameters, is a high priority (see Chapters 10–14 of this book).

5   The soil carbon stocks are temperature sensitive, and MRT also depends on the carbon types (brown, white, blue, black, green and frozen C) and their origins and locations in addition to soil inherent parameters such as texture and mineralogy.

6   Blue carbon, the ecosystem C stock in coastal ecosystems, is large, highly valuable and the most threatened by changes due to human activities. Yet, mangroves and other coastal ecosystems provide a range of ecosystem services, including C sequestration for the mitigation of climate change. The world's most C-rich mangroves contain 703 ± 38 Mg C/ha.

7   Global hot spots of C stocks are permafrost, tropical rainforest, coastal ecosystems, wetlands/peat lands and urban lands. Sustainable management strategies to maintain and foster the soil carbon pools in these environments need to be adopted urgently.

8   Technological options for the management of soil health through the restoration of SOC stocks are conservation agriculture, integrated nutrient management, agroforestry, landscape management and improved grazing systems, which are described in more detail in the following chapters.

## 17 Future trends in research

Additional research is needed, for site-specific conditions, with regards to the following:

1.  Critical limits of soil organic carbon concentration in the root zone in relation to optimization of agronomic yield and maximization of the use efficiency of inputs (e.g. fertilizers, irrigation, improved varieties, pesticide use).

2.  Incremental increase in agronomic yield with unit increase in soil organic carbon concentration or stock in the root zone.

3.  Saving of inputs (e.g. fertilizer, irrigation, energy) with unit increase in soil organic carbon stock in the root zone.

4.  In-field measurements of soil organic carbon concentration by cost-effective and easy-to-use hand-held devices.

5. Assessment of the rate of sequestration of soil inorganic carbon (e.g. secondary carbonates in soil of arid/semi-arid regions and in irrigated ecosystems.
6. Estimates of societal value of soil organic carbon for the purpose of payments to farmers and trading of carbon credits.
7. Identification and implementation of policies which promote and incentivize farmers and land manager for adoption of practices which lead to re-carbonization of soils and of the terrestrial biosphere.
8. Promoting curricula at school and levels which enhance awareness among students about the importance of soil are source or sink of atmosphere $CO_2$.

# 18 References

Alvarez, R., Berhongaray, G. and Gimenez, A. (2021). Are grassland soils of the pampas sequestering carbon? *Science of the Total Environment* 763: 142978. DOI: 10.1016/j.scitotenv.2020.142978.

Arrouays, D., Saby, N., Walter, C., Lemercier, B. and Schvartz, C. (2006). Relationships between particle-size distribution and organic carbon in French arable topsoils. *Soil Use and Management* 22(1): 48–51. DOI: 10.1111/j.1475-2743.2006.00020.x.

Atwood, T. B., Witt, A., Mayorga, J., Hammill, E. and Sala, E. (2020). Global patterns in marine sediment carbon stocks. *Frontiers in Marine Science* 7. DOI: 10.3389/fmars.2020.00165.

Bai, J., Yu, L., Du, S., Wei, Z., Liu, Y., Zhang, L., Zhang, G. and Wang, X. (2020). Effects of flooding frequencies on soil carbon and nitrogen stocks in river marginal wetlands in a ten-year period. *Journal of Environmental Management* 267: 110618. DOI: 10.1016/j.jenvman.2020.110618.

Balesdent, J. and Recous, S. (1997). Les temps de résidence du carbone et le potentiel de stockage de carbone dans quelques sols cultivés français. *Canadian Journal of Soil Science* 77(2): 187–193. DOI: 10.4141/S96-109.

Batjes, N. H. (1996). Total carbon and nitrogen in the soils of the world. *European Journal of Soil Science*. Blackwell Publishing Ltd 47(2): 151–163. DOI: 10.1111/j.1365-2389.1996.tb01386.x.

Bork, E. W., Raatz, L. L., Carlyle, C. N., Hewins, D. B. and Thompson, K. A. (2020). Soil carbon increases with long-term cattle stocking in northern temperate grasslands. *Soil Use and Management* 36(3): 387–399. DOI: 10.1111/sum.12580.

Bradley-Cook, J. I. and Virginia, R. A. (2018). Landscape variation in soil carbon stocks and respiration in an Arctic tundra ecosystem, west Greenland. *Arctic, Antarctic, and Alpine Research* 50(1). DOI: 10.1080/15230430.2017.1420283.

Burke, L., Kura, Y., Kassem, K., Revenga, C., Spalding, M. and McAllister, D. (2001). *Pilot Analysis of Global Ecosystems: Agroecosystems*. Washington, DC: World Resources Institute. Available at: http://pdf.wri.org/page_coastal.pdf.

Cao, M. and Woodward, F. I. (1998). Net primary and ecosystem production and carbon stocks of terrestrial ecosystems and their responses to climate change. *Global Change Biology* 4(2): 185–198. DOI: 10.1046/j.1365-2486.1998.00125.x.

Chatterjee, A. (2020). Soil carbon pools of six ecological regions of the United States. *Journal of Forestry Research* 31(5): 1933–1938. DOI: 10.1007/s11676-019-00976-z.

Chen, S., Huang, Y., Zou, J. and Shi, Y. (2013). Mean residence time of global topsoil organic carbon depends on temperature, precipitation and soil nitrogen. *Global and Planetary Change* 100: 99–108. DOI: 10.1016/j.gloplacha.2012.10.006.

Cotrufo, M. F., Ranalli, M. G., Haddix, M. L., Six, J. and Lugato, E. (2019). Soil carbon storage informed by particulate and mineral-associated organic matter. *Nature Geoscience* 12(12): 989–994. DOI: 10.1038/s41561-019-0484-6.

Crowther, T. W., Riggs, C., Lind, E. M., Borer, E. T., Seabloom, E. W., Hobbie, S. E., Wubs, J., Adler, P. B., Firn, J., Gherardi, L., Hagenah, N., Hofmockel, K. S., Knops, J. M. H., McCulley, R. L., MacDougall, A. S., Peri, P. L., Prober, S. M., Stevens, C. J. and Routh, D. (2019). Sensitivity of global soil carbon stocks to combined nutrient enrichment. *Ecology Letters* 22(6): 936–945. DOI: 10.1111/ele.13258.

Crutzen, P. J. and Stoermer, E. F. (2000). The anthropocene, global change. *International Geosphere-Biosphere Programme (IGBP)* 41: 17–18.

Dantas, D., Terra, MdC. N. S. de, Pinto, L. O. R., Calegario, N. and Maciel, S. M. (2021). Above and belowground carbon stock in a tropical forest in Brazil. *Acta Scientiarum. Agronomy* 43: e48276. DOI: 10.4025/actasciagron.v43i1.48276.

de Carvalho, A. Md, Marchão, R. L., Souza, K. W. and Bustamante, M. MdC. (2014). Soil fertility status, carbon and nitrogen stocks under cover crops and tillage regimes. *Revista Ciencia Agronomica* 45(5spe): 914–921. DOI: 10.1590/S1806-66902014000500007.

Devi, S. B. and Sherpa, S. S. S. S. (2019). Soil carbon and nitrogen stocks along the altitudinal gradient of the Darjeeling Himalayas, India. *Environmental Monitoring and Assessment* 191(6): 361. DOI: 10.1007/s10661-019-7470-8.

Djagbletey, E. D., Logah, V., Ewusi-Mensah, N. and Tuffour, H. O. (2018). Carbon stocks in the Guinea savanna of Ghana: estimates from three protected areas. *Biotropica* 50(2): 225–233. DOI: 10.1111/btp.12529.

Dungait, J. A. J., Hopkins, D. W., Gregory, A. S. and Whitmore, A. P. (2012). Soil organic matter turnover is governed by accessibility not recalcitrance. *Global Change Biology*. John Wiley & Sons, Ltd 18(6): 1781–1796. DOI: 10.1111/j.1365-2486.2012.02665.x.

Eswaran, H., Van Den Berg, E. and Reich, P. (1993). Organic carbon in soils of the world. *Soil Science Society of America Journal* 57(1): 192–194. DOI: 10.2136/sssaj1993.03615995005700010034x.

Fu, C., Li, Y., Zeng, L., Zhang, H., Tu, C., Zhou, Q., Xiong, K., Wu, J., Duarte, C. M., Christie, P. and Luo, Y. (2021). Stocks and losses of soil organic carbon from Chinese vegetated coastal habitats. *Global Change Biology* 27(1): 202–214. DOI: 10.1111/gcb.15348.

Gajić, A., Ramke, H. G., Hendricks, A. and Koch, H. (2012). Microcosm study on the decomposability of hydrochars in a cambisol. *Biomass and Bioenergy* 47: 250–259. DOI: 10.1016/j.biombioe.2012.09.036.

Garten, C. T. and Ashwood, T. L. (2002). Landscape level differences in soil carbon and nitrogen: implications for soil carbon sequestration. *Global Biogeochemical Cycles: 61-1-61–14* 16(4): 61–61. DOI: 10.1029/2002GB001918.

Gruba, P., Socha, J., Błońska, E., Lasota, J., Suchanek, A. and Gołąb, P. (2014). Influence of parent material on the spatial distribution of organic carbon stock in the forest soils. *Sylwan* 158(6): 443–452.

Herr, D. and Landis, E. (2016). *Coastal Blue Carbon Ecosystems. Opportunities for Nationally Determined Contributions. Policy Brief.* Gland, Switzerland and Washington, DC: IUCN and TNC.

Himes, F. L. (1998). Nitrogen, sulfur, and phosphorus and the sequestering of carbon. In: Lal, R., Kimble, J. M., Follett, R. F., et al. (Eds) *Soil Processes and the Carbon Cycle*. Boca Raton, FL: CRC Press, pp. 315–319. DOI: 10.1201/9780203739273.

Hou, C., Li, Y., Huang, Y., Zhu, H., Ma, J., Yu, F. and Zhang, X. (2021). Reclamation substantially increases soil organic and inorganic carbon stock in riparian floodplains. *Journal of Soils and Sediments* 21(2): 957–966. DOI: 10.1007/s11368-020-02836-4.

Huang, J., Minasny, B., McBratney, A. B., Padarian, J. and Triantafilis, J. (2018). The location- and scale- specific correlation between temperature and soil carbon sequestration across the globe. *Science of the Total Environment* 615: 540–548. DOI: 10.1016/j.scitotenv.2017.09.136.

Huntzinger, D. N., Schaefer, K., Schwalm, C., Fisher, J. B., Hayes, D., Stofferahn, E., Carey, J., Michalak, A. M., Wei, Y., Jain, A. K., Kolus, H., Mao, J., Poulter, B., Shi, X., Tang, J. and Tian, H. (2020). Evaluation of simulated soil carbon dynamics in Arctic-Boreal ecosystems. *Environmental Research Letters* 15(2). DOI: 10.1088/1748-9326/ab6784.

IPCC (2013). 2013: the physical science basis. Contribution of working group I to the fifth assessment report of the Intergovernmental Panel on Climate Change. In Stocker, T. F., Qin, D., Plattner, G.-K., et al. (Eds). *Climate Change*. Cambridge and New York: CNY. Cambridge University Press.

IUCN (2021). *Blue Carbon*. Available at: https://www.iucn.org/resources/issues-briefs/blue-carbon.

Jardine, S. L. and Siikamäki, J. V. (2014). A global predictive model of carbon in mangrove soils. *Environmental Research Letters* 9(10). DOI: 10.1088/1748-9326/9/10/104013.

Jhariya, M. K. and Singh, L. (2021). Effect of fire severity on soil properties in a seasonally dry forest ecosystem of Central India. *International Journal of Environmental Science and Technology* 18(12): 3967–3978. DOI: 10.1007/s13762-020-03062-8.

Jones, C., McConnell, C., Coleman, K., Cox, P., Falloon, P., Jenkinson, D. and Powlson, D. (2005). Global climate change and soil carbon stocks; predictions from two contrasting models for the turnover of organic carbon in soil. *Global Change Biology* 11(1): 154–166. DOI: 10.1111/j.1365-2486.2004.00885.x.

Jones, I. L., DeWalt, S. J., Lopez, O. R., Bunnefeld, L., Pattison, Z. and Dent, D. H. (2019). Above- and belowground carbon stocks are decoupled in secondary tropical forests and are positively related to forest age and soil nutrients respectively. *Science of the Total Environment* 697: 133987. DOI: 10.1016/j.scitotenv.2019.133987.

Jungkunst, H. F., Krüger, J. P., Heitkamp, F., et al. (2012). Accounting more precisely for peat and other soil carbon resources. In: Lal, R., Lorenz, K., Hüttl, R. F., et al. (Eds) *Recarbonization of the Biosphere: Ecosystem and the Global Carbon Cycle*. Dordrecht, Holland: Springer, pp. 127–157.

Kassam, A., Friedrich, T. and Derpsch, R. (2019). Global spread of Conservation Agriculture. *International Journal of Environmental Studies*. Routledge 76(1): 29–51. DOI: 10.1080/00207233.2018.1494927.

Kauffman, J. B., Bernardino, A. F., Ferreira, T. O., Giovannoni, L. R., de O. Gomes, L. E., Romero, D. J., Jimenez, L. C. Z. and Ruiz, F. (2018). Carbon stocks of mangroves and salt marshes of the Amazon region, Brazil. *Biology Letters* 14(9). DOI: 10.1098/rsbl.2018.0208.

Kauffman, J. B., Giovanonni, L., Kelly, J., Dunstan, N., Borde, A., Diefenderfer, H., Cornu, C., Janousek, C., Apple, J. and Brophy, L. (2020). Total ecosystem carbon stocks at the marine-terrestrial interface: blue carbon of the Pacific Northwest Coast, United States. *Global Change Biology* 26(10): 5679–5692. DOI: 10.1111/gcb.15248.

Köchy, M., Hiederer, R. and Freibauer, A. (2015). Global distribution of soil organic carbon - Part 1: masses and frequency distributions of SOC stocks for the tropics, permafrost regions, wetlands, and the world. *SOIL* 1(1): 351–365. DOI: 10.5194/soil-1-351-2015.

Krull, E. S., Baldock, J. A. and Skjemstad, J. O. (2003). Importance of mechanisms and processes of the stabilisation of soil organic matter for modelling carbon turnover. *Functional Plant Biology* 30(2): 207–222. DOI: 10.1071/FP02085.

Kuznetsova, A. I., Lukina, N. V., Gornov, A. V., Gornova, M. V., Tikhonova, E. V., Smirnov, V. E., Danilova, M. A., Tebenkova, D. N., Braslavskaya, T. Y., Kuznetsov, V. A., Tkachenko, Y. N. and Genikova, N. V. (2020). Carbon stock in sandy soils of pine forests in the west of Russia. *Eurasian Soil Science* 53(8): 1056–1065. DOI: 10.1134/S1064229320080104.

Lal, R. (2004). Soil carbon sequestration impacts on global climate change and food security. *Science* 304(5677): 1623–1627. DOI: 10.1126/science.1097396.

Lal, R. (2015). A system approach to conservation agriculture. *Journal of Soil and Water Conservation* 70(4): 82A–88A. DOI: 10.2489/jswc.70.4.82A.

Lal, R. (2018). Digging deeper: a holistic perspective of factors affecting soil organic carbon sequestration in agroecosystems. *Global Change Biology*. Blackwell Publishing Ltd 24(8): 3285–3301. DOI: 10.1111/gcb.14054.

Lal, R. (2019). Carbon cycling in global drylands. *Current Climate Change Reports*. Springer 5(3): 221–232. DOI: 10.1007/s40641-019-00132-z.

Larionova, A. A., Zolotareva, B. N., Yevdokimov, I. V., Sapronov, D. V., Kuzyakov, Y. V. and Buegger, F. (2008). The rates of organic matter renewal in gray forest soils and chernozems. *Eurasian Soil Science* 41(13): 1378–1386. DOI: 10.1134/S106422930813005X.

Larionova, A. A., Stulin, A. F., Zanina, O. G., Yevdokimov, I. V., Khokhlova, O. S., Buegger, F., Schloter, M. and Kudeyarov, V. N. (2012). Distribution of stable carbon isotopes in an agrochernozem during the transition from C3 vegetation to a corn monoculture. *Eurasian Soil Science* 45(8): 768–778. DOI: 10.1134/S1064229312060063.

Lehndorff, E., Roth, P. J., Cao, Z. H. and Amelung, W. (2014). Black carbon accrual during 2000 years of paddy-rice and non-paddy cropping in the Yangtze River Delta, China. *Global Change Biology* 20(6): 1968–1978. DOI: 10.1111/gcb.12468.

Leifeld, J., Bassin, S. and Fuhrer, J. (2005). Carbon stocks in Swiss agricultural soils predicted by land-use, soil characteristics, and altitude. *Agriculture, Ecosystems and Environment* 105(1–2): 255–266. DOI: 10.1016/j.agee.2004.03.006.

Leifeld, J., Bassin, S., Conen, F., Hajdas, I., Egli, M. and Fuhrer, J. (2013). Control of soil pH on turnover of belowground organic matter in subalpine grassland. *Biogeochemistry* 112(1–3): 59–69. DOI: 10.1007/s10533-011-9689-5.

Leite, L. F. C., Mendonça, E. S., Neves, J. C. L., Machado, P. L. O. A. and Galvão, J. C. C. (2003). Estoques totais de carbono orgânico e seus compartimentos em argissolo sob floresta e sob milho cultivado com adubação mineral e orgânica. *Revista Brasileira de Ciência do Solo* 27(5): 821–832. DOI: 10.1590/S0100-06832003000500006.

Lim, S. S., Baah-Acheamfour, M., Choi, W. J., Arshad, M. A., Fatemi, F., Banerjee, S., Carlyle, C. N., Bork, E. W., Park, H. and Chang, S. X. (2018). Soil organic carbon stocks in three Canadian agroforestry systems: from surface organic to deeper mineral soils. *Forest Ecology and Management* 417: 103–109. DOI: 10.1016/j.foreco.2018.02.050.

Lugato, E., Lavallee, J. M., Haddix, M. L., Panagos, P. and Cotrufo, M. F. (2021). Different climate sensitivity of particulate and mineral-associated soil organic matter. *Nature Geoscience* 14(5): 295–300. DOI: 10.1038/s41561-021-00744-x.

Lutfalla, S., Abiven, S., Barré, P., Wiedemeier, D. B., Christensen, B. T., Houot, S., Kätterer, T., Macdonald, A. J., van Oort, F. and Chenu, C. (2017). Pyrogenic carbon lacks long-term persistence in temperate arable soils. *Frontiers in Earth Science* 5. DOI: 10.3389/feart.2017.00096.

Machado, P. L. OdA. (2005). Soil carbon and the mitigation of global climate change. *Quimica Nova* 28(2): 329–334. DOI: 10.1590/S0100-40422005000200026.

Macreadie, P. I., Anton, A., Raven, J. A., Beaumont, N., Connolly, R. M., Friess, D. A., Kelleway, J. J., Kennedy, H., Kuwae, T., Lavery, P. S., Lovelock, C. E., Smale, D. A., Apostolaki, E. T., Atwood, T. B., Baldock, J., Bianchi, T. S., Chmura, G. L., Eyre, B. D., Fourqurean, J. W., Hall-Spencer, J. M., Huxham, M., Hendriks, I. E., Krause-Jensen, D., Laffoley, D., Luisetti, T., Marbà, N., Masque, P., McGlathery, K. J., Megonigal, J. P., Murdiyarso, D., Russell, B. D., Santos, R., Serrano, O., Silliman, B. R., Watanabe, K. and Duarte, C. M. (2019). The future of blue carbon science. *Nature Communications* 10(1): 3998. DOI: 10.1038/s41467-019-11693-w.

McNicol, G., Bulmer, C., D'Amore, D., Sanborn, P., Saunders, S., Giesbrecht, I., Arriola, S. G., Bidlack, A., Butman, D. and Buma, B. (2019). Large, climate-sensitive soil carbon stocks mapped with pedology-informed machine learning in the North Pacific coastal temperate rainforest. *Environmental Research Letters* 14(1). DOI: 10.1088/1748-9326/aaed52.

Mikhailova, E. A. and Post, C. J. (2006). Effects of land use on soil inorganic carbon stocks in the Russian chernozem. *Journal of Environmental Quality* 35(4): 1384–1388. DOI: 10.2134/jeq2005.0151.

Mishra, U., Hugelius, G., Shelef, E., Yang, Y., Strauss, J., Lupachev, A., Harden, J. W., Jastrow, J. D., Ping, C. L., Riley, W. J., Schuur, E. A. G., Matamala, R., Siewert, M., Nave, L. E., Koven, C. D., Fuchs, M., Palmtag, J., Kuhry, P., Treat, C. C., Zubrzycki, S., Hoffman, F. M., Elberling, B., Camill, P., Veremeeva, A. and Orr, A. (2021). Spatial heterogeneity and environmental predictors of permafrost region soil organic carbon stocks. *Science Advances* 7(9): eaaz5236. DOI: 10.1126/sciadv.aaz5236.

Mitsch, W. J. and Gosselink, J. G. (2015). Wetlands. 5th ed. Hoboken, NJ.

Moharana, P. C., Jena, R. K., Kumar, N., Singh, R. S. and Rao, S. S. (2021). Assessment of soil organic and inorganic carbon stock at different soil depths after conversion of desert into arable land in the hot arid regions of India. *Carbon Management* 12(2): 153–165. DOI: 10.1080/17583004.2021.1893128.

Nehren, U. and Wicaksono, P. (2018). Mapping soil carbon stocks in an oceanic mangrove ecosystem in Karimunjawa Islands, Indonesia. *Estuarine, Coastal and Shelf Science* 214: 185–193. DOI: 10.1016/j.ecss.2018.09.022.

Nguyen, B. T., Koide, R. T., Dell, C., Drohan, P., Skinner, H., Adler, P. R. and Nord, A. (2014). Turnover of soil carbon following addition of switchgrass-derived biochar to four soils. *Soil Science Society of America Journal* 78(2): 531–537. DOI: 10.2136/sssaj2013.07.0258.

Okpoho, N. A. (2018). Smallholder agriculture land use impact on soil organic carbon stock in Federal Capital Territory of Nigeria. *Journal of Agriculture and Environment for International Development* 112(1): 109–119. DOI: 10.12895/jaeid.20181.702.

Ouillon, S. (2018). Why and how do we study sediment transport? Focus on coastal zones and ongoing methods. *Water* 10(4): 390. DOI: 10.3390/w10040390.

Ouyang, X. and Lee, S. Y. (2020). Improved estimates on global carbon stock and carbon pools in tidal wetlands. *Nature Communications* 11(1): 317. DOI: 10.1038/s41467-019-14120-2.

Parshotam, A., Saggar, S., Searle, P. L., Daly, B. K., Sparling, G. P. and Parfitt, R. L. (2000). Carbon residence times obtained from labelled ryegrass decomposition in soils under contrasting environmental conditions. *Soil Biology and Biochemistry* 32(1): 75-83. DOI: 10.1016/S0038-0717(99)00131-5.

Paul, S., Flessa, H., Veldkamp, E. and López-Ulloa, M. (2008). Stabilization of recent soil carbon in the humid tropics following land use changes: evidence from aggregate fractionation and stable isotope analyses. *Biogeochemistry* 87(3): 247-263. DOI: 10.1007/s10533-008-9182-y.

Polat, A., Chim, B., Kumar, S. and Osborne, S. (2020). On-farm assessment of soil quality in low and high grazing under integrated crop-livestock system in south Dakota. *Tarim Bilimleri Dergisi* 26(4): 434-441. DOI: 10.15832/ankutbd.557832.

Rasse, D. P., Rumpel, C. and Dignac, M. F. (2005). Is soil carbon mostly root carbon? Mechanisms for a specific stabilisation. *Plant and Soil* 269(1-2): 341-356. DOI: 10.1007/s11104-004-0907-y.

Reithmaier, G. M. S., Johnston, S. G., Junginger, T., Goddard, M. M., Sanders, C. J., Hutley, L. B., Ho, D. T. and Maher, D. T. (2021). Alkalinity production coupled to pyrite formation represents an unaccounted blue carbon sink. *Global Biogeochemical Cycles* 35(4). DOI: 10.1029/2020GB006785.

Reyna-Bowen, L., Lasota, J., Vera-Montenegro, L., Vera-Montenegro, B. and Błońska, E. (2019). Distribution and factors influencing organic carbon stock in mountain soils in Babia Góra National Park, Poland. *Applied Sciences (Switzerland)* 9(15). DOI: 10.3390/app9153070.

Ricart, A. M., York, P. H., Rasheed, M. A., Pérez, M., Romero, J., Bryant, C. V. and Macreadie, P. I. (2015). Variability of sedimentary organic carbon in patchy seagrass landscapes. *Marine Pollution Bulletin* 100(1): 476-482. DOI: 10.1016/j.marpolbul.2015.09.032.

Ruddiman, W. F. (2003). The anthropogenic greenhouse era began thousands of years ago. *Climatic Change* 61(3): 261-293. DOI: 10.1023/B:CLIM.0000004577.17928.

Rumpel, C., Dignac, M. F., Remuscat, L. and Barré, P. (2015). Nanoscale evidence of contrasted process for root-derived organic matter stabilization by mineral interactions depending on soil depth. *Soil Biology and Biochemistry*. https://doi.org/10.1016/j.soilbio.2015.02.017.

Sanderman, J. (2012). Can management induced changes in the carbonate system drive soil carbon sequestration? A review with particular focus on Australia. *Agriculture, Ecosystems and Environment* 155: 70-77. DOI: 10.1016/j.agee.2012.04.015.

Sanderman, J. and Amundson, R. (2008). A comparative study of dissolved organic carbon transport and stabilization in California forest and grassland soils. *Biogeochemistry* 89(3): 309-327. DOI: 10.1007/s10533-008-9221-8.

Sanderman, J. and Amundson, R. (2009). A comparative study of dissolved organic carbon transport and stabilization in California forest and grassland soils. *Biogeochemistry* 92(1-2): 41-59. DOI: 10.1007/s10533-008-9249-9.

Sanderman, J., Amundson, R. G. and Baldocchi, D. D. (2003). Application of eddy covariance measurements to the temperature dependence of soil organic matter mean residence time. *Global Biogeochemical Cycles* 17(2). DOI: 10.1029/2001GB001833.

Scharlemann, J. P. W., Tanner, E. V. J., Hiederer, R. and Kapos, V. (2014). Global soil carbon: understanding and managing the largest terrestrial carbon pool. *Carbon Management*. Taylor & Francis 5(1): 81-91. DOI: 10.4155/cmt.13.77.

Setia, R., Gottschalk, P., Smith, P., Marschner, P., Baldock, J., Setia, D. and Smith, J. (2013). Soil salinity decreases global soil organic carbon stocks. *Science of the Total Environment* 465: 267–272. DOI: 10.1016/j.scitotenv.2012.08.028.

Shi, H. J., Wang, X. J., Zhao, Y. J., Xu, M. G., Li, D. W. and Guo, Y. (2017). Relationship between soil inorganic carbon and organic carbon in the wheat-maize cropland of the North China Plain. *Plant and Soil* 418(1–2): 423–436. DOI: 10.1007/s11104-017-3310-1.

Shyamsundar, P., Springer, N. P., Tallis, H., Polasky, S., Jat, M. L., Sidhu, H. S., Krishnapriya, P. P., Skiba, N., Ginn, W., Ahuja, V., Cummins, J., Datta, I., Dholakia, H. H., Dixon, J., Gerard, B., Gupta, R., Hellmann, J., Jadhav, A., Jat, H. S., Keil, A., Ladha, J. K., Lopez-Ridaura, S., Nandrajog, S. P., Paul, S., Ritter, A., Sharma, P. C., Singh, R., Singh, D. and Somanathan, R. (2019). Fields on fire: alternatives to crop residue burning in India. *Science* 365(6453): 536–538. DOI: 10.1126/science.aaw4085.

Sierra, C. A., Hoyt, A. M., He, Y. and Trumbore, S. E. (2018). Soil organic matter persistence as a stochastic process: age and transit time distributions of carbon in soils. *Global Biogeochemical Cycles* 32(10): 1574–1588. DOI: 10.1029/2018GB005950.

Six, J., Conant, R. T., Paul, E. A. and Paustian, K. (2002). Stabilization mechanisms of soil organic matter: implications for C-saturation of soils. *Plant and Soil* 241(2): 155–176. DOI: 10.1023/A:1016125726789.

Smith, S. W., Vandenberghe, C., Hastings, A., Johnson, D., Pakeman, R. J., van der Wal, R. and Woodin, S. J. (2014). Optimizing carbon storage Within a spatially heterogeneous upland grassland Through sheep grazing management. *Ecosystems* 17(3): 418–429. DOI: 10.1007/s10021-013-9731-7.

Spohn, M. (2020). Increasing the organic carbon stocks in mineral soils sequesters large amounts of phosphorus. *Global Change Biology* 26(8): 4169–4177. DOI: 10.1111/gcb.15154.

Street, L. E., Mielke, N. and Woodin, S. J. (2018). Phosphorus availability determines the response of tundra ecosystem carbon stocks to nitrogen enrichment. *Ecosystems* 21(6): 1155–1167. DOI: 10.1007/s10021-017-0209-x.

Tan, W. F., Zhang, R., Cao, H., Huang, C., Yang, Q., Wang, M. and Koopal, L. K. (2014). Soil inorganic carbon stock under different soil types and land uses on the Loess Plateau region of China. *CATENA* 121: 22–30. DOI: 10.1016/j.catena.2014.04.014.

Tanwar, S. P. S., Kumar, P., Verma, A., Bhatt, R. K., Singh, A., Lal, K., Patidar, M. and Mathur, B. K. (2019). Carbon sequestration potential of agroforestry systems in the Indian arid zone. *Current Science* 117(12): 2014–2022. DOI: 10.18520/cs/v117/i12/2014-2022.

The Blue Carbon Initiative (2019). *Mitigating Climate Change through Coastal Ecosystem Management.* Conservation International, IUCN, Intergovernmental Oceanographic Cooperation. Available at: https://www.thebluecarboninitiative.org/.

Tifafi, M., Guenet, B. and Hatté, C. (2018). Large differences in global and regional total soil carbon stock estimates based on SoilGrids, HWSD, and NCSCD: intercomparison and evaluation based on field data From USA, England, wales, and France. *Global Biogeochemical Cycles.* John Wiley & Sons, Ltd 32(1): 42–56. DOI: 10.1002/2017GB005678.

Tomar, U. and Baishya, R. (2020). Moisture regime influence on soil carbon stock and carbon sequestration rates in semi-arid forests of the National Capital Region, India. *Journal of Forestry Research* 31(6): 2323–2332. DOI: 10.1007/s11676-019-01032-6.

Trumbore, S. and de Camargo, P. B. (2009). Soil carbon dynamics. In: *Geophysical Monograph Series.* Keller, M., Bustamante, M. and Gash, J. (Eds) Geophysical Monograph Series, pp. 451–462. DOI: 10.1029/2008GM000741.

van der Kamp, J., Yassir, I. and Buurman, P. (2009). Soil carbon changes upon secondary succession in Imperata grasslands (east Kalimantan, Indonesia). *Geoderma* 149(1-2): 76–83. DOI: 10.1016/j.geoderma.2008.11.033.

Vasenev, V. and Kuzyakov, Y. (2018). Urban soils as hot spots of anthropogenic carbon accumulation: review of stocks, mechanisms and driving factors. *Land Degradation and Development* 29(6): 1607–1622. DOI: 10.1002/ldr.2944.

Vasenev, V. I., Stoorvogel, J. J. and Vasenev, I. I. (2013). Urban soil organic carbon and its spatial heterogeneity in comparison with natural and agricultural areas in the Moscow region. *CATENA* 107: 96–102. DOI: 10.1016/j.catena.2013.02.009.

Wäldchen, J., Schulze, E. D., Schöning, I., Schrumpf, M. and Sierra, C. (2013). The influence of changes in forest management over the past 200 years on present soil organic carbon stocks. *Forest Ecology and Management* 289: 243–254. DOI: 10.1016/j.foreco.2012.10.014.

Walker, X. J., Baltzer, J. L., Cumming, S. G., Day, N. J., Ebert, C., Goetz, S., Johnstone, J. F., Potter, S., Rogers, B. M., Schuur, E. A. G., Turetsky, M. R. and Mack, M. C. (2019) Increasing wildfires threaten historic carbon sink of boreal forest soils. *Nature* 572(7770): 520–523. doi: 10.1038/s41586-019-1474-y. Epub 2019 Aug 21. PMID: 31435055.

Wang, J., Tang, J., Li, Z., Yang, W., Yang, P. and Qu, Y. (2020). Corn and rice cultivation affect soil organic and inorganic carbon storage through altering soil properties in alkali sodic soils, Northeast of China. *Sustainability* 12(4). DOI: 10.3390/su12041627.

Wattel-Koekkoek, E. J. W., Buurman, P., Van Der Plicht, J., Wattel, E. and Van Breemen, N. (2003). Mean residence time of soil organic matter associated with kaolinite and smectite. *European Journal of Soil Science* 54(2): 269–278. DOI: 10.1046/j.1365-2389.2003.00512.x.

Yang, D., Miao, X. Y., Wang, B., Jiang, R. P., Wen, T., Liu, M. S., Huang, C. and Xu, C. (2020). System-specific complex interactions shape soil organic carbon distribution in coastal salt marshes. *International Journal of Environmental Research and Public Health* 17(6). DOI: 10.3390/ijerph17062037.

Yang, Y., Fang, J., Ji, C., Ma, W., Mohammat, A., Wang, S., Wang, S., Datta, A., Robinson, D. and Smith, P. (2012). Widespread decreases in topsoil inorganic carbon stocks across China's grasslands during 1980s-2000s. *Global Change Biology* 18(12): 3672–3680. DOI: 10.1111/gcb.12025.

Yu, W., Ding, X., Xue, S., Li, S., Liao, X. and Wang, R. (2013). Effects of organic-matter application on phosphorus adsorption of three soil parent materials. *Journal of Soil Science and Plant Nutrition* 13(ahead): 0. DOI: 10.4067/S0718-95162013005000079.

Yu, X., Zhou, W., Wang, Y., Cheng, P., Hou, Y., Xiong, X., Du, H., Yang, L. and Wang, Y. (2020). Effects of land use and cultivation time on soil organic and inorganic carbon storage in deep soils. *Journal of Geographical Sciences* 30(6): 921–934. DOI: 10.1007/s11442-020-1762-3.

Zamanian, K., Zhou, J. and Kuzyakov, Y. (2021). Soil carbonates: the unaccounted, irrecoverable carbon source. *Geoderma* 384. DOI: 10.1016/j.geoderma.2020.114817.

Zeraatpisheh, M. and Khormali, F. (2013). Carbon stock and mineral factors controlling soil organic carbon in a climatic gradient, Golestan province. *Journal of Soil Science and Plant Nutrition* 12: 637–654. https://doi.org/10.4067/S0718-95162012005000022.

Zhang, K., Wang, X., Wu, L., Lu, T., Guo, Y. and Ding, X. (2021). Impacts of salinity on the stability of soil organic carbon in the croplands of the Yellow River Delta. *Land Degradation and Development* 32(4): 1873–1882. DOI: 10.1002/ldr.3840.

# Chapter 3

## Measuring and monitoring soil carbon sequestration

*Matthias Kuhnert, Sylvia H. Vetter and Pete Smith, Institute of Biological & Environmental Sciences, University of Aberdeen, UK*

## 1 Introduction

Soils have recently received attention in the context of climate change, because they are the largest terrestrial carbon pool (Batjes, 1996, Lal, 2004), and small changes in this large reservoir may affect atmospheric $CO_2$ concentrations. To preserve current atmospheric $CO_2$ levels and to limit climate change impacts, it is thus important to protect the carbon stored in soils and prevent its release into the atmosphere. Carbon may be stored in soils at long timescales, and soil carbon sequestration may therefore have a large potential as negative emission technology (Paustian et al., 2016, Minasny et al., 2017, Paustian et al., 2019). Such technologies need to be employed to meet the climate targets of the Paris Agreement or the different national net zero targets (Climate Ambition Alliance: Net Zero 2050). Increasing soil organic carbon (SOC) sequestration will remove atmospheric carbon and store it in the soil. This process can have a range of positive side effects (Smith et al., 2021). Rumpel et al. (2018) describe eight steps to make the soil more resilient and more productive and improve the storage of carbon. One central part of awarding land managers for improving soils and for the application of SOC storage as a large-scale climate mitigation strategy is the availability of a system for measuring, reporting and verification (MRV) of the effect of land management practice. This includes the quantification of SOC changes over time. However, SOC sequestration is a complex and slow process affected by a wide range of factors (see Chapters 2–9 of this book). This makes the

http://dx.doi.org/10.19103/AS.2022.0106.09

measurement and monitoring of SOC challenging. While point measurements are associated with errors and uncertainties, large-scale quantification presents even more challenges, as it requires either large amounts of samples, which are costly and labour intensive, or upscaling methods based on assumptions.

Soils are used for different purposes and are often managed by a variety of users with their own interests. Activities to increase SOC are not necessarily the main interest of the users and/or landowners. Farmers, for example, need to rely on constant harvest to provide food and make a living. Incentives or legal obligations are required to introduce changes to maintain and/or increase SOC (see Chapters 26–29 of this book), and their implementation depends on a functional monitoring system. This system will be based on the available and future development of tools and approaches to quantify SOC changes (measurements, modelling, etc.). Available tools show a wide range of complexity and accuracy, with a general trend towards the application of simpler options with easy, cheap and rapid methods.

To account for varying complexity of methodologies, the Intergovernmental Panel on Climate Change (IPCC) introduced a three-step tier system, with Tier 1 indicating a basic method with an equation and default emission factors, Tier 2 using the same equation but country-/region-specific emission factors and Tier 3 any more complex method, ranging from alternative equations to process-based models. The equation and default factor for Tier 1 describes a linear relation between an activity (e.g. fertilizer application on the field) and the related estimated greenhouse gas (GHG) emissions. Scientific data build the basis for this relation, which is a simplified but effective and standardised approach to estimating GHG emissions. While the used values in Tier 1 are more generic (based on global data), the Tier 2 approach is similar to Tier 1 but uses country-specific values that provide a more accurate estimate for the target country. Available emission factors are summarised in the emission factor database (https://www.ipcc-nggip.iges.or.jp/EFDB/main.php).

There is an increasing interest in the economics of carbon sequestration. While there is already an interest in an investment in more sustainable companies (Kareiva et al., 2015), carbon accounting and trading of carbon units generated by SOC sequestration has started already. However, all of these different interests rely on a functional and applicable MRV system. Recent research focused on MRV applications and has developed suitable frameworks (Paustian et al., 2019, Smith et al., 2020, FAO, 2020). In this chapter, we will present available systems and discuss their advantages and limitations.

## 2 Measurement/monitoring, reporting and verification

In this chapter, we distinguish MRV frameworks from MRV applications. While a framework provides a more theoretical description of an optimum MRV system, the application describes an applied MRV system, using protocols that include concrete definitions of used models, measurement approaches and other required details (e.g. responsibilities for the different actions). Smith et al. (2020) outlined a generic concept for an MRV framework as a combination of different approaches to quantify SOC change over time. Generally, management changes are applied on a field or farm level. The farm level is the scale for the beneficiaries of subsidies or carbon trading. Therefore, field and farm scales are most relevant to an MRV scheme. A central part of quantifying SOC changes on this scale is field measurements, but these are costly and labour intensive. Additionally, MRV protocols often lack clear measurement standards (Bispo et al., 2017). While some aspects are clarified (depth of the top 30 cm, 1 m if possible; FAO, 2020), other specifications are missing (number of samples, date of sampling relative to the management practices, spatial distribution of sampling, etc.). Therefore, alternative approaches to measurements need to be considered. Modelling is a very attractive alternative, as all problems and limitations of the measurements are resolved by using a model. But the quality of the simulation result needs to be considered, especially in comparison to measurements. Models include errors and uncertainty based on the assumptions used and the underlying concepts, and they require data for calibration and validation, in addition to those needed for running the models. MRV frameworks, as outlined by the FAO (2020) and Smith et al. (2020), specify that only calibrated models can contribute to SOC quantification, but further specification of the models is not provided. A combination of both (measurement and modelling) will compensate for the disadvantages of each other and improve the result (Smith et al., 2020). Overall, a combination of different approaches secures the optimum quantification of SOC changes over time. Smith et al. (2020) list seven components of an MRV framework: long-term experimental sites, field experiments (short term), field-specific modelling, spatial data analysis combined with modelling, collection and aggregation of activity data (e.g. conventional and intervention management), remote sensing and spatial re-sampling. The different components complement each other to allow an optimum framework for measuring and verifying SOC changes over time. There are advantages and disadvantages of all different components, and all show some limitations. Only a combination of all, or at least several of these methods, will provide good MRV outcomes.

## 2.1 Measurements

Direct measurements of SOC content involve quantifying the fine earth and coarse earth fraction, the organic carbon concentration and soil bulk density or fine earth mass (FAO, 2019). Estimating the rock content of sample soils can be a challenge but will significantly affect soil bulk density (Poeplau et al.,

2017, Throop et al., 2012). Another challenge is that a change in management (whole practice as well as depth at which that practice is applied) will not only impact the bulk density of the soil but also the amount of soil in a soil sample at a certain depth (Haynes and Naidu, 1998). Therefore, corrections and use of the equivalent mass approach may be necessary (Chapter 11 of this book). As soils are characterised by high spatial variability, direct measurements rely on appropriate study designs and sampling protocols (Minasny et al., 2017, Chapter 12 of this book). At the field scale, a large number of soil samples are usually required to give reliable SOC stock estimates with an acceptable error margin (Garten and Wullschleger, 1999, Vanguelova et al., 2016).

The IPCC recommends a sample depth of 30 cm, but several methods for increasing SOC content require deeper sampling to confirm the expected effect (Smith et al., 2020). For example, the effect of a no-tillage practice on the SOC content may be overestimated if the measuring depth is insufficient (Angers and Eriksen-Hamel, 2008, Blanco-Canqui and Lal, 2008).

A change in SOC stocks can also be estimated through indirect measurements and by presenting the full carbon budget. This approach uses the net balance of carbon fluxes measured through chamber measurements or the eddy covariance (EC) method (Baldocchi, 2003). From the carbon fluxes, the initial uptake of carbon through photosynthesis and its subsequent partial loss through respiration (from soil, plant and litter) are estimated to give net ecosystem exchange or net ecosystem production and further C inputs (organic fertilization) and outputs (harvest) to and from the system (Smith et al., 2010, Soussana et al., 2010). Through this complied carbon budget, a change in SOC can be estimated. This approach indirectly measures the change in SOC for larger landscapes but can only be used under horizontal homogeneity of the footprint area and under sufficient air turbulences (Aubinet et al., 1999). The maintenance of most measurement systems is costly and time-consuming. The post-processing of the measured data also needs time and expert knowledge about flux corrections for density and gap filling (Falge et al., 2001, Reichstein et al., 2005). In an MRV application, EC provides landscape-specific data, which can be used as baseline data or for model optimisation purposes (calibration and validation).

Long-term study sites are crucial for the implementation of the MRV framework (Smith et al., 2020). Study sites for different management combinations allow a long-term observation and quantification of all relevant parameters and variables that affect SOC sequestration. 'Long-term' is relative and not defined as a fixed duration. The IPCC suggests 20 years as the default period to observe SOC changes because SOC sequestration rates are fast at the beginning but slow down over time until they approach zero (Sommer and Bossio, 2014). Measurements are impractical for a generic implementation of an MRV process (too costly and labour intensive), and other solutions that replace or at least reduce the sampling intensity in the field are required. Besides modelling and remote sensing, long-term study sites in combination with short-term field experiments can complement field measurements. These study sites provide data for re-assessment of potential impacts, reference for expected changes or baseline for a particular management practice. Additionally, these data will be the basis for the development, calibration and validation of models and remote sensing approaches. Ideally, land cover, soil, climate, management and environmental conditions are represented by available study sites or at least a reasonable number of combinations (good representation of all climate zones, soil types, crop species, etc.). A standard protocol for the acquisition of these data would be beneficial, because differences in the set-up of the measurement approaches could introduce uncertainty. Organizing and providing these data on accessible platforms is the best way for an open and transparent handling of the data. Two platforms for long-term experimental sites were initiated by the SOMNET (Smith et al., 2002) and the EuroSOMNET (https://www.ufz.de/somnet/, Franko et al., 2002) platforms. The SOMNET platform evolved later into an online, real-time inventory project, including a webpage with Long-Term Soil-Ecosystems Experiments. The database contains meta-data of more than 200 long-term experiments and is hosted by the International Soil Carbon Network (http://iscn.fluxdata.org/network/partner-networks/ltse/). More than 80% of the long-term experimental sites concern agricultural systems (Smith et al., 2012). However, the majority of the sites are in the temperate climate zone with a focus on Europe and North America, under-representing tropical and sub-tropical regions and the Southern hemisphere (Smith et al., 2012). For good coverage of the variability of global agricultural systems, more long-term sites in other parts of the world need to be established. The better the representation of different management options, soil and climate zones by experimental sites, the better the data basis for MRV application. This requires immediate action, as study sites that are established today, will be able to be used to assess long-term effects on SOC in 20 years (Smith et al., 2012). Special funding is required to initiate long-term monitoring sites, as project funding for 3-5 years duration is insufficient.

## 2.2 Remote sensing

Besides *in-situ* measurements, remote sensing can support the monitoring of SOC changes and/or provide data for the verification of measured SOC changes. This technology allows non-invasive measurements, including at a large scale. Remote sensing can be applied in the lab or in the field by handheld or transportable systems or by airborne or satellite device devices (Chabrillat et al., 2019). As part of an MRV application, the latter two options are more useful, as these systems allow a wider coverage and delivery of large-scale data globally. Considering the wide application and availability of the data, this would reduce costs for monitoring SOC changes (Nocita et al., 2015) once the approach is established. There are different approaches used for SOC estimation, and two of them are highlighted below.

One established remote sensing approach is reflectance spectroscopy. It uses characteristic spectra that are reflected from the soil surface for quantitative and quantitative analysis of soil properties. The recommended wavelength range for these measurements is the visible near infrared–shortwave infrared (700–2500 nm), as it shows a good signal-to-noise ratio and is a cost- and time-effective option for spectroscopy (Mohamed et al., 2018). The characteristic spectra are reflected by the bonds in the SOC molecules (O–H, N–H, C–H), which allow a qualitative and quantitative analysis of SOC. This method provides soil-type-specific quantitative SOC estimates (Grinand et al., 2012). To secure a wider application without site specific measurements, spectral libraries are required that contain several thousand soil types with varying soil properties as a reference. This is a cost- and time-effective alternative to other traditional measurement options in the laboratory, such as wet digestion or dry combustion (Nayak et al., 2019).

The introduced high spectroscopy measures for fixed wavelength using multispectral sensors, which is associated with some limitations, especially for quantitative measurements on SOC (Ben-Dor et al., 2018). Therefore, recent developments in hyperspectral sensors show an improved approach with higher capability for quantitative data over large areas. Hyperspectral remote sensing (also called image spectroscopy) provides a continuous spectrum for each pixel, using 100 or more contiguous spectral bands. However, Ben-Dor et al. (2018) also list a wide range of drawbacks with the signal-to-noise ratio as a major problem (caused, e.g. by the non-transparent atmosphere, problems with sensor calibration) and more problems with changing conditions (e.g. changes in soil particle size). Further developments are required to improve the approach. For large-scale applications, there is again a demand for developing new libraries for the new approach.

The advantage of large-scale remote sensing using airborne devices and/or satellites is that it can provide additional information, e.g. land-use change (Winkler et al., 2021), primary production (Zhao et al., 2005) or different soil properties (Viscarra Rossel et al., 2006). Nevertheless, remote sensing has limitations. The availability of images is affected by cloud cover, measurements are affected by plant cover on the ground and only the top centimetre can be measured (Smith et al., 2020). Despite good results in different studies and the availability of spectral libraries, the measurement is still uncertain, which renders remote sensing as the sole MRV method unsuitable. In contrast, it is an excellent additional approach to complement other methods and should therefore be used only in combination.

The latest developments in multispectral systems to quantify SOC have shown great progress (Aldana-Jague et al., 2016). These kinds of measurements have the potential to reduce uncertainty (Chabrillat et al., 2019), new libraries have to be built for the new approach. Chabrillat et al. (2019) refer also to studies using hyperspectral systems (Gomez et al., 2008, Lu et al., 2013) but rates the performance as moderate. Similar to the other approaches, remote sensing shows good potential to complement measurements and reduced costs but is not able to replace field measurements completely.

## 2.3 Modelling

As there are limitations to field measurements and remote sensing, modelling becomes the most prominent supplement to provide data for MRV applications. Models can contribute in different ways to MRV: (1) provide baseline information, (2) interpolate measurements (temporally and spatially), (3) extrapolate measurements for projections or for an ex-ante assessment, (4) estimate SOC changes, and (5) provide information for an optimised measurement plan. Different models with different complexity and accuracy can be used in MRV applications. However, there are no standards defining the quality of a model used in an MRV application. Choosing the right model depends on the objective, data availability and modelling skills of the user (Table 1), as different models vary in their characteristics, complexity and accuracy (Table 1).

Biogeochemical models seem to be most suitable for MRV approaches, as they are able to simulate SOC with the highest accuracy and provide additional information about impacts on yield and GHG emissions (Camino-Serrano et al., 2018, Campbell and Paustian, 2015). However, these models are sometimes impractical, as they require a large amount of data and expert knowledge to use them. Emission factors and simple equations are often used by carbon trading platforms, but these models are developed for large-scale (country scale) applications and might show large errors on the field scale. Most suitable seems to be decision support tools for carbon accounting, like the Cool Farm Tool (CFT, https://coolfarmtool.org/, Hillier et al., 2011) or COMET-Farm (http://comet-farm.com/, Paustian et al., 2017), as they address or show the potential to address the aforementioned problems of the emission factors and process-based models (Whittaker et al., 2013). These tools use different routines of different complexities (Tier 1 to Tier 3, depending on the tool). Unfortunately, the most popular options also show limitations, as the SOC component of the CFT uses the Tier 1 approach of the IPCC 2006 guidelines (although this is under review at the moment) and the COMET-Farm Tool uses a Tier 3 approach in combination with a database that only covers the United States. Further developments of both tools are ongoing, and in the future, these may be reasonable options. Both tools are developed for stakeholders with

**Table 1** Specifications for different model categories

|  | Emission factors | Decision support tools | SOC models | Biogeochemical models (Tier 3) |
|---|---|---|---|---|
| Data requirement | Low | High (farm-specific data) | High (environmental data) | High (environmental data) |
| Calibration requirement | Low | Low | High | High |
| Required expertise | Low | Medium | High | High |
| Management options | Medium (categories) | Medium-high | No-high | High |
| Targeted scale | Country and larger | Field-farm | Point/site | Point/site |
| Uncertainty/expected error for field scale | High | Medium-high | Low | Low |
| Example models | UNFCCC models, Tier 1, Tier 2 | Cool Farm Tool, Comet Farm Tool | RothC | EPIC, DAYCENT, DNDC |

The category emission factors also include simple empirical equations (Tier 2 approaches). The categories SOC and biogeochemical models (Tier 3 approaches) are separated to indicate if a model only includes SOC dynamics or also addresses processes (e.g. N cycle, plant growth). The category decision support includes tools that are designed to provide information on GHG emissions, SOC changes or both (these tools use mainly Tier 1 and Tier 2 approaches but can also include Tier 3 routines).

an easy-to-use interface. The CFT was developed with an interface usable by farmers and the input information they have at hand. The CFT calculates the GHG emissions on a farm level for a specific site and specific management. The methods used within CFT range from emission factors to a model approach considering region-specific parameters and farm-level data (e.g. management, soil, climate) on an annual basis. Therefore, the CFT can be seen as a Tier 2 or simple Tier 3 model. COMET-Farm calculates the carbon footprint of a farm. The tool provides the opportunity to test different management interventions and explore their mitigation potential, i.e. the potential reduction of GHG emissions. GHG estimates for crops are calculated using the DayCent dynamic model (Del Grosso et al., 2010, Parton et al., 1998) – a process-based model – and follow the official USDA GHG inventory guidelines for entity-scale reporting (Eve et al., 2014). Both tools consider soil carbon sequestration and calculate the SOC change for a land use change or change in soil management.

## 3 Existing monitoring, reporting and verification protocols

The FAO (2020) published a protocol that provides concrete guidelines about the structure and steps to apply an MRV application (Fig. 1). The FAO protocol differentiates reporting into four different categories: (1) pre-implementation report, (2) initial report, (3) biannual report and (4) final report. These reports describe the different stages of MRV frameworks as outlined in Fig. 1. The protocol also provides suggestions and guidelines for the responsibilities of MRV framework. It is suggested that the reporting can be organised by the farmer but needs to be done in consultation with a relevant expert. An independent person or entity must verify the reports. The monitoring and verification require expert knowledge, which can be secured by the accreditation of independent experts specialised in these activities and by external expert reviews, respectively. The accreditation and certification can be organised by governmental institutions (e.g. for subsidies) and other large organisations (e.g. the FAO or entities). Certain specifications are not included (e.g. measurement approaches, suitable models) as the protocol is a blueprint for global use and might require local adaptation.

In addition to the FAO's standardised framework, two other examples of MRV protocols in Alberta, Canada and Australia are already in place. The Government of Alberta published an MRV protocol to quantify the impacts of tillage management on GHG emissions and SOC. It differs from the FAO protocol by specifying the target area and the management to be applied, which allows some aspects to be considered in more detail. For example, some reversal events are allowed for natural farms. Conventional tillage is allowed in less than 10% of the farm area for weed control. Another protocol has been published by

**Stage S1: Applicability conditions**

-restrict non-suitable areas and options
-estimate and proof potential benefits
-restrict leakage (offset outside boundaries)

**Stage S2: Delineating boundaries**

-Spatial boundaries
-Locations
-Temporal boundaries

**Stage S3: Delineate baseline and intervention scenarios**

-Define business as usual scenario
-The management of the last 5 years is baseline
-Define intervention scenario

**Stage S4: Preliminary assessment of SOC and GHG emissions**

-Is the sequestering practice additional?
-What is the 'benchmark' farming practice?
-How much impact is additional?

**Stage S5: Monitoring**

-Soil sample monitoring
-SOC modelling
-GHG estimates monitoring

**Stage S6: Reporting and verification**

-Reporting by four report types
-Verification needs to be independent of monitoring and reporting

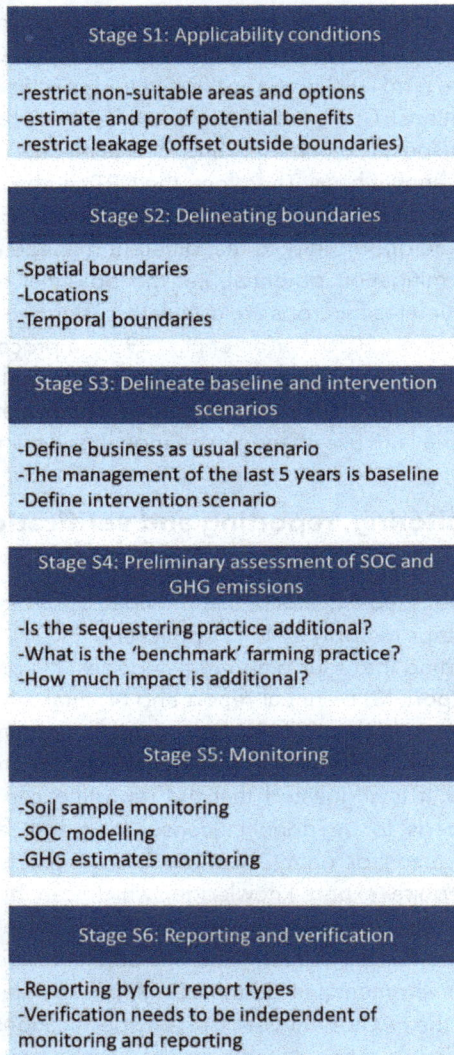

**Figure 1** The MRV framework of the FAO (2020) follows a six-stage approach to set up an MRV protocol.

the Australian Government (Smith et al., 2020), which includes bare land and pastures alongside croplands for baseline conditions. This protocol has a similar structure and content as the FAO framework, but it is more specific in defining some details in the management options that are allowed but differs in some other details (e.g. review every 5 years).

Carbon accounting platforms have also started to trade carbon based on SOC gains. The good standards of the protocols are undermined by their implementation. One example of the actual protocols is the Verified Carbon Standard (Shoch et al., 2020). Measurements are very limited (one measurement suggested), SOC changes are quantified by Tier 1 models and the project time is restricted to a short period (e.g. 10 years). The implementation of a simple MRV application by businesses is economically motivated. Even though there is a demand for cost reduction, the methods applied need to be improved to provide an adequate data for carbon trading. Nevertheless, improvements in this sector would provide a business solution that will improve mitigation actions once a functional system is established.

## 4 Outlook of the use of monitoring, reporting and verification applications

MRV applications are essential to the implementation of strategies to mitigate climate change (Smith et al., 2020). It is also a requirement for subsidies from governmental institutions, carbon trading or the initiatives of companies with net zero targets (FAO, 2020, Kareiva et al., 2015, Paustian et al., 2019). In contrast to MRV applications for other processes or variables, monitoring of SOC stock changes has additional challenges: (1) the slow rate of change in soil carbon against the large background stock, (2) the heterogeneous distribution in space and depth and (3) the complexity of measurements and the reversibility of the gains, which make the requirements of MRV complex. MRV protocols overcome these problems by applying a combination of different methods, to compensate for the limitations of individual quantification methods. The implementation of MRV applications requires an integrated approach but barriers exist. The combined approach can be costly, labour intensive and/or requires a wider skill set (or even expert knowledge). In summary, current MRV methods are often impractical for stakeholders.

In the near future, the challenge for science is to reduce complexity and to remove these barriers in order to provide practical solutions. One relatively easy target could be the development or improvement of models that are easily applicable by stakeholders but that provide robust results at the field and farm scale. More challenging and more time intensive will be the further development of remote sensing approaches. Remote sensing will never be able to replace field measurements, but it will improve the quality of the measurements and might allow for a reduction of the number of samples to save labour time and lower costs.

Other approaches like digital mapping will also contribute to an improved understanding and quantification of SOC changes. Such developments have the potential to improve the measurements in MRV applications.

The following chapters will further detail some aspects of MRV approaches. Chapter 11 of this book will give an overview of methods for quantifying SOC stocks and characterising its turnover times at the profile scale. Chapter 12 of this book will introduce digital soil mapping as an additional option to quantify SOC on a farm level (De Gruijter et al., 2015). The chapter will indicate the advantages and limitations of this approach, including the measurement demand and the associated uncertainty. Chapter 13 of this book will give a detailed overview of SOC modelling approaches with a special focus on its permanency. Finally, Chapter 14 of this book will outline digital stock taking, with a focus on the field scale. This will include an analysis of knowledge gaps in field-specific digital stock taking and new approaches, such as the application of smartphones to quantify SOC stocks. These methods will be discussed in the context of an application in MRV applications, which would bring down measurement costs and potentially improve accuracy.

## 5 Conclusion

Sequestering atmospheric carbon through increased SOC stocks requires a functional MRV application to monitor the impacts of management practices on the soil. A single quantification approach is not sufficient; instead a combination of different methods is necessary to monitor SOC changes over time and to provide appropriate verification methods. For simplified MRV applications, there is a risk of errors and uncertainty. Developments of the available tools do not meet all the demands for an MRV application applied for different purposes. There is an imbalance between complexity and accuracy (for modelling) as well as in costs and accuracy (measurements). The chapters in this section describe currently available approaches and future developments that might provide effective solutions to be applied in MRV applications. The approaches presented do not target the implementation in MRV applications, but they are measurement tools, which can be used in MRV applications.

## 6 Where to look for further information

The carbon trading market for soil organic carbon developed very recently. Therefore, there are no agreed standards for MRV systems. The number of trading platforms is increasing, and the protocols of the existing platforms are diverse and still changing. These platforms usually provide the MRV protocols as reports on their webpages, which provide detailed information about the accounting process. However, as these are businesses, no platform should be highlighted here. In a very recent publication, different accounting protocols were compared by Oldfield et al. (2022).

As mentioned, there are no agreed standards and policymakers acknowledged this gap. Therefore, there are a couple of calls for upcoming projects (starting 2023) that will tackle this problem (e.g. projects funded by the HORIZON program of the European Union). These projects as well as some ongoing projects (e.g. ClieNFarms (European), RETINA (UK)) will provide information on this topic in future. The results of former projects and existing governmental initiatives on this topic are listed in the list of references and introduced in the text.

Beside the key publication by Smith et al. (2020), the Food and Agricultural Organization of the United Nations provides a global standard for the application of an MRV system (https://www.fao.org/documents/card/en/c/cb0509en/).

The European Soil Data Centre published in 2015 a sampling protocol to verify changes of the organic carbon stock in the soil for a region in Italy (https://esdac.jrc.ec.europa.eu/content/validation-eu-soil-sampling-protocol -verify-changes-organic-carbon-stock-mineral-soil). Additionally, the European Commission had recently published a working document on Sustainable Carbon Cycles – Carbon Farming (https://ec.europa.eu/clima/system/files /2021-12/swd_2021_450_en_0.pdf).

# 7 References

Aldana-Jague, E., Heckrath, G., Macdonald, A., van Wesemael, B. and Van Oost, K. 2016. UAS-based soil carbon mapping using VIS-NIR (480-1000 nm) multi-spectral imaging: potential and limitations, *Geoderma* 275, 55-66.

Angers, D. A. and Eriksen-Hamel, N. S. 2008. Full-inversion tillage and organic carbon distribution in soil profiles: a meta-analysis, *Soil Science Society of America Journal* 72(5), 1370-1374.

Aubinet, M., Grelle, A., Ibrom, A., Rannik, Ü., Moncrieff, J., Foken, T., Kowalski, A. S., Martin, P. H., Berbigier, P., Bernhofer, C., Clement, R., Elbers, J., Granier, A., Grünwald, T., Morgenstern, K., Pilegaard, K., Rebmann, C., Snijders, W., Valentini, R. and Vesala, T. 1999. Estimates of the annual net carbon and water exchange of forests: the EUROFLUX methodology, *Advances in Ecological Research* 30, 113-175.

Baldocchi, D. D. 2003. Assessing the eddy covariance technique for evaluating carbon dioxide exchange rates of ecosystems: past, present and future, *Global Change Biology* 9(4), 479-492.

Batjes, N. H. 1996. Total carbon and nitrogen in the soils of the world, *European Journal of Soil Science* 47(2), 151-163.

Ben-Dor, E., Chabrillat, S. and Dematte, J. A. M. 2018. Characterization of soil properties using reflectance spectroscopy. In: *Hyperspectral Remote Sensing of Vegetation* (2nd edn.), Thenkabail, P., Lyon, J. and Huete, A. (Eds), CRC Press, Boca Raton, pp. 187-161.

Bispo, A., Andersen, L., Angers, D. A., Bernoux, M., Brossard, M., Cecillon, L., Comans, R. N. J., Harmsen, J., Jonassen, K., Lame, F., Lhuillery, C., Maly, S., Martin, E., Mcelnea,

A. E., Sakai, H., Watabe, Y. and Eglin, T. K. 2017. Accounting for carbon stocks in soils and measuring GHGs emission fluxes from soils: do we have the necessary standards?, *Frontiers in Environmental Science* 5, 41.

Blanco-Canqui, H. and Lal, R. 2008. No-tillage and soil-profile carbon sequestration: an on-farm assessment, *Soil Science Society of America Journal* 72(3), 693–701.

Camino-Serrano, M., Guenet, B., Luyssaert, S., Ciais, P., Bastrikov, V., De Vos, B., Gielen, B., Gleixner, G., Jornet-Puig, A., Kaiser, K., Kothawala, D., Lauerwald, R., Penuelas, J., Schrumpf, M., Vicca, S., Vuichard, N., Walmsley, D. and Janssens, I. A. 2018. ORCHIDEE-SOM: modeling soil organic carbon (SOC) and dissolved organic carbon (DOC) dynamics along vertical soil profiles in Europe, *Geoscientific Model Development* 11(3), 937–957.

Campbell, E. E. and Paustian, K. 2015. Current developments in soil organic matter modeling and the expansion of model applications: a review, *Environmental Research Letters* 10(12), 123004.

Chabrillat, S., Ben-Dor, E., Cierniewski, J., Gomez, C., Schmid, T. and van Wesemael, B. 2019. Imaging spectroscopy for soil mapping and monitoring, *Surveys in Geophysics* 40(3), 361–399.

Climate Ambition Alliance: Net Zero 2050 2021. Available at: https://climateaction.unfccc.int/views/cooperative-initiative-details.html?id=94.

De Gruijter, J. J., Minasny, B. and McBratney, A. B. 2015. Optimizing stratification and allocation for design-based estimation of spatial means using predictions with error, *Journal of Survey Statistics and Methodology* 3(1, March), 19–42.

Del Grosso, S. J., Ogle, S. M., Parton, W. J. and Breidt, F. J. 2010. Estimating uncertainty in $N_2O$ emissions from US cropland soils, *Global Biogeochemical Cycles* 24(1), GB1009.

Eve, M., Flugge, M. and Pape, D. 2014. Chapter 2: considerations when estimating agriculture and forestry GHG emissions and removals. In: *Quantifying Greenhouse Gas Fluxes in Agriculture and Forestry: Methods for Entity-Scale Inventory*, Eve, M., Pape, D., Flugge, M., et al. (Eds), Office of the Chief Economist, U.S. Department of Agriculture, Washington, DC, pp. 2-1-2-25.

Falge, E., Baldocchi, D., Olson, R., Anthoni, P., Aubinet, M., Bernhofer, C., Burba, G., Ceulemans, R., Clement, R., Dolman, H., Granier, A., Gross, P., Grunwald, T., Hollinger, D., Jensen, N. O., Katul, G., Keronen, P., Kowalski, A., Lai, C. T., Law, B. E., Meyers, T., Moncrieff, J., Moors, E., Munger, J. W., Pilegaard, K., Rannik, Ü., Rebmann, C., Suyker, A., Tenhunen, J., Tu, K., Verma, S., Vesala, T., Wilson, K. and Wofsy, S. 2001. Gap filling strategies for defensible annual sums of net ecosystem exchange, *Agricultural and Forest Meteorology* 107(1), 43–69.

FAO. 2020. *A Protocol for Measurement, Monitoring, Reporting and Verification of Soil Organic Carbon in Agricultural Landscapes-GSOC MRV Protocol*, Food and Agriculture Organization of the United Nations, Rome.

FAO. 2019. *Measuring and Modelling Soil Carbon Stocks and Stock Changes in Livestock Production Systems: Guidelines for Assessment (1st version). Livestock Environmental Assessment and Performance (LEAP) Partnership*, FAO, Rome, Italy.

Franko, U., Schramm, G., Rodionova, V., Korschens, M., Smith, P., Coleman, K., Romanenkov, V. and Shevtsova, L. 2002. EuroSOMNET - a database for long-term experiments on soil organic matter in Europe, *Computers and Electronics in Agriculture* 33(3), 233–239.

Garten, C. T. and Wullschleger, S. D. 1999. Soil carbon inventories under a bioenergy crop (switchgrass): measurement limitations, *Journal of Environmental Quality* 28(4), 1359–1365.

Gomez, C., Rossel, R. A. V. and McBratney, A. B. 2008. Soil organic carbon prediction by hyperspectral remote sensing and field vis-NIR spectroscopy: an Australian case study, *Geoderma* 146(3-4), 403-411.

Grinand, C., Barthes, B. G., Brunet, D., Kouakoua, E., Arrouays, D., Jolivet, C., Caria, G. and Bernoux, M. 2012. Prediction of soil organic and inorganic carbon contents at a national scale (France) using mid-infrared reflectance spectroscopy (MIRS), *European Journal of Soil Science* 63(2), 141-151.

Haynes, R. J. and Naidu, R. 1998. Influence of lime, fertilizer and manure applications on soil organic matter content and soil physical conditions: a review, *Nutrient Cycling in Agroecosystems* 51(2), 123-137.

Hillier, J., Walter, C., Malin, D., Garcia-Suarez, T., Mila-i-Canals, L. and Smith, P. 2011. A farm-focused calculator for emissions from crop and livestock production, *Environmental Modelling and Software* 26(9), 1070-1078.

Kareiva, P. M., McNally, B. W., McCormick, S., Miller, T. and Ruckelshaus, M. 2015. Improving global environmental management with standard corporate reporting, *Proceedings of the National Academy of Sciences of the United States of America* 112(24), 7375-7382.

Lal, R. 2004. Soil carbon sequestration impacts on global climate change and food security, *Science* 304(5677), 1623-1627.

Lu, P., Wang, L., Niu, Z., Li, L. and Zhang, W. 2013. Prediction of soil properties using laboratory VIS-NIR spectroscopy and Hyperion imagery, *Journal of Geochemical Exploration* 132, 26-33.

Minasny, B., Malone, B. P., McBratney, A. B., Angers, D. A., Arrouays, D., Chambers, A., Chaplot, V., Chen, Z., Cheng, K., Das, B. S., Field, D. J., Gimona, A., Hedley, C. B., Hong, S. Y., Mandal, B., Marchant, B. P., Martin, M., McConkey, B. G., Mulder, V. L., O'Rourke, S., Richer-de-Forges, A. C., Odeh, I., Padarian, J., Paustian, K., Pan, G., Poggio, L., Savin, I., Stolbovoy, V., Stockmann, U., Sulaeman, Y., Tsui, C., Vågen, T., van Wesemael, B. and Winowiecki, L. 2017. Soil carbon 4 per mille, *Geoderma* 292, 59-86.

Mohamed, E. S., Saleh, A. M., Belal, A. B. and Gad, A. 2018. Application of near-infrared reflectance for quantitative assessment of soil properties, *Egyptian Journal of Remote Sensing and Space Science* 21(1), 1-14.

Nayak, A. K., Rahman, M. M., Naidu, R., Dhal, B., Swain, C. K., Nayak, A. D., Tripathi, R., Shahid, M., Islam, M. R. and Pathak, H. 2019. Current and emerging methodologies for estimating carbon sequestration in agricultural soils: a review, *Science of the Total Environment* 665, 890-912.

Nocita, M., Stevens, A., van Wesemael, B., Aitkenhead, M., Bachmann, M., Barthès, B., Ben Dor, E., Brown, D. J., Clairotte, M., Csorba, A., Dardenne, P., Demattê, J. A. M., Genot, V., Guerrero, C., Knadel, M., Montanarella, L., Noon, C., Ramirez-Lopez, L., Robertson, J., Sakai, H., Soriano-Disla, J. M., Shepherd, K. D., Stenberg, B., Towett, E. K., Vargas, R. and Wetterlind, J. 2015. Soil spectroscopy: an alternative to wet chemistry for soil monitoring, *Advances in Agronomy* 132, 139-159.

Oldfield, E. E., Eagle, A. J., Rubin, R. L., Rudek, J., Sanderman, J., Gordon, D. R. 2022. Crediting agricultural soil carbon sequestration, *Science* 375(6586), 1222-1225. doi: 10.1126/science.abl7991. Epub 2022 Mar 17. PMID: 35298251.

Parton, W. J., Hartman, M., Ojima, D. and Schimel, D. 1998. DAYCENT and its land surface submodel: description and testing, *Global and Planetary Change* 19(1-4), 35-48.

Paustian, K., Collier, S., Baldock, J., Burgess, R., Creque, J., DeLonge, M., Dungait, J., Ellert, B., Frank, S., Goddard, T., Govaerts, B., Grundy, M., Henning, M., Izaurralde, R.

C., Madaras, M., McConkey, B., Porzig, E., Rice, C., Searle, R., Seavy, N., Skalsky, R., Mulhern, W. and Jahn, M. 2019. Quantifying carbon for agricultural soil management: from the current status toward a global soil information system, *Carbon Management* 10(6), 567–587.

Paustian, K., Easter, M., Brown, K., Chambers, A., Eve, M., Huber, A., Marx, E., Layer, M., Stermer, M., Sutton, B., Swan, A., Toureene, C., Verlayudhan, S. and Williams, S. 2017. Field- and farm-scale assessment of soil greenhouse gas mitigation using COMET-Farm, *Agronomy Monographs*, 341–359.

Paustian, K., Lehmann, J., Ogle, S., Reay, D., Robertson, G. P. and Smith, P. 2016. Climate-smart soils, *Nature* 532(7597), 49–57.

Poeplau, C., Vos, C. and Don, A. 2017. Soil organic carbon stocks are systematically overestimated by misuse of the parameters bulk density and rock fragment content, *SOIL* 3(1), 61–66.

Reichstein, M., Falge, E., Baldocchi, D., Papale, D., Aubinet, M., Berbigier, P., Bernhofer, C., Buchmann, N., Gilmanov, T., Granier, A., Grunwald, T., Havrankova, K., Ilvesniemi, H., Janous, D., Knohl, A., Laurila, T., Lohila, A., Loustau, D., Matteucci, G., Meyers, T., Miglietta, F., Ourcival, J. M., Pumpanen, J., Rambal, S., Rotenberg, E., Sanz, M., Tenhunen, J., Seufert, G., Vaccari, F., Vesala, T., Yakir, D. and Valentini, R. 2005. On the separation of net ecosystem exchange into assimilation and ecosystem respiration: review and improved algorithm, *Global Change Biology* 11(9), 1424–1439.

Rumpel, C., Amiraslani, F., Koutika, L. S., Smith, P., Whitehead, D. and Wollenberg, E. 2018. Put more carbon in soils to meet Paris climate pledges, *Nature* 564(7734), 32–34.

Shoch, D., Swails, E. and indigo 2020. *Methodology for Improved Agricultural Land Management*. Verified, Carbon Standard. VCS Methodology, version 1. Report VW0042.

Smith, P., Davies, C. A., Ogle, S., Zanchi, G., Bellarby, J., Bird, N., Boddey, R. M., McNamara, N. P., Powlson, D., Cowie, A., Noordwijk, M., Davis, S. C., Richter, D. D. B., Kryzanowski, L., Wijk, M. T., Stuart, J., Kirton, A., Eggar, D., Newton-Cross, G., Adhya, T. K. and Braimoh, A. K. 2012. Towards an integrated global framework to assess the impacts of land use and management change on soil carbon: current capability and future vision, *Global Change Biology* 18(7), 2089–2101.

Smith, P., Falloon, P., Korschens, M., Shevtsova, L., Franko, U., Romanenkov, V., Coleman, K., Rodionova, V., Smith, J. and Schramm, G. 2002. EuroSOMNET - a European database of long-term experiments on soil organic matter: the WWW metadatabase, *Journal of Agricultural Science* 138(2), 123–134.

Smith, P., Keesstra, S. D., Silver, W. L., Adhya, T. K., De Deyn, G. B., Carvalheiro, L. G., Giltrap, D. L., Renforth, P., Cheng, K., Sarkar, B., Saco, P. M., Scow, K., Smith, J. U., Morel, J. C., Thiele-Bruhn, S., Lal, R. and McElwee, P. 2021. Soil-derived nature's contributions to people and their contribution to the UN sustainable development goals, *Philosophical Transactions of the Royal Society of London. Series B, Biological Sciences* 376(1834), 20200185.

Smith, P., Lanigan, G., Kutsch, W. L., Buchmann, N., Eugster, W., Aubinet, M., Ceschia, E., Beziat, P., Yeluripati, J. B., Osborne, B., Moors, E. J., Brut, A., Wattenbach, M., Saunders, M. and Jones, M. 2010. Measurements necessary for assessing the net ecosystem carbon budget of croplands, *Agriculture, Ecosystems and Environment* 139(3), 302–315.

Smith, P., Soussana, J. F., Angers, D., Schipper, L., Chenu, C., Rasse, D. P., Batjes, N. H., van Egmond, F., McNeill, S., Kuhnert, M., Arias-Navarro, C., Olesen, J. E., Chirinda,

N., Fornara, D., Wollenberg, E., Álvaro-Fuentes, J., Sanz-Cobena, A. and Klumpp, K. 2020. How to measure, report and verify soil carbon change to realize the potential of soil carbon sequestration for atmospheric greenhouse gas removal, *Global Change Biology* 26(1), 219-241.

Sommer, R. and Bossio, D. 2014. Dynamics and climate change mitigation potential of soil organic carbon sequestration, *Journal of Environmental Management* 144, 83-87.

Soussana, J. F., Tallec, T. and Blanfort, V. 2010. Mitigating the greenhouse gas balance of ruminant production systems through carbon sequestration in grasslands, *Animal: An International Journal of Animal Bioscience* 4(3), 334-350.

Throop, H. L., Archer, S. R., Monger, H. C. and Waltman, S. 2012. When bulk density methods matter: implications for estimating soil organic carbon pools in rocky soils, *Journal of Arid Environments* 77, 66-71.

Vanguelova, E. I., Bonifacio, E., De Vos, B., Hoosbeek, M. R., Berger, T. W., Vesterdal, L., Armolaitis, K., Celi, L., Dinca, L., Kjonaas, O. J., Pavlenda, P., Pumpanen, J., Puttsepp, Ü., Reidy, B., Simoncic, P., Tobin, B. and Zhiyanski, M. 2016. Sources of errors and uncertainties in the assessment of forest soil carbon stocks at different scales-review and recommendations, *Environmental Monitoring and Assessment* 188(11), 630.

Viscarra Rossel, R. A., Walvoort, D. J. J., McBratney, A. B., Janik, L. J. and Skjemstad, J. O. 2006. Visible, near infrared, mid infrared or combined diffuse reflectance spectroscopy for simultaneous assessment of various soil properties, *Geoderma* 131(1-2), 59-75.

Whittaker, C., McManus, M. C. and Smith, P. 2013. A comparison of carbon accounting tools for arable crops in the United Kingdom, *Environmental Modelling and Software* 46, 228-239.

Winkler, K., Fuchs, R., Rounsevell, M. and Herold, M. 2021. Global land use changes are four times greater than previously estimated, *Nature Communications* 12(1), 2501.

Zhao, M., Heinsch, F. A., Nemani, R. R. and Running, S. W. 2005. Improvements of the MODIS terrestrial gross and net primary production global data set, *Remote Sensing of Environment* 95(2), 164-176.

# Chapter 4

## Spectral mapping of soil organic carbon

*Bas van Wesemael, Université catholique de Louvain, Belgium*

## 1 Introduction

Croplands have lost the majority of their soil organic carbon since the onset of agriculture (Sanderman *et al.*, 2017). The main reasons for this loss are the reduced return of plant residues to the soil compared to forest and grassland systems as well as the increase in microbial activity resulting from the disturbance by tillage, harrowing, fertilization and harvesting (Chenu *et al.*, 2019). Soil organic matter and its main component soil organic carbon (SOC) are essential for preserving a healthy soil and maintaining soil fertility. It is proposed that soil health is dependent on the maintenance of four major functions: carbon transformations; nutrient cycles; soil structure maintenance and the regulation of pests and diseases (Kibblewhite *et al.*, 2008). A meta-analysis by Oldfield *et al.* (2019) showed that SOC content was related to yield potential of wheat and maize and that an increase in SOC concentrations up to region-specific targets can reduce reliance on nitrogen fertilizer and help close the yield gap. Moreover, Minasny *et al.* (2017) demonstrated the potential of atmospheric $CO_2$ sequestration through increase in SOC stock of a range of agricultural soils along a global climate gradient. Unfortunately, high-resolution maps that indicate areas where SOC levels are critically low and where improved management options, such as cover crops, agroforestry no-till, could be most effective are scarce. Recent SOC maps at regional to global scale consist of rather coarse grid cells that are too coarse to represent heterogeneity at the field scale and are based on legacy data that is generally not up-to-date as

http://dx.doi.org/10.19103/AS.2020.0079.18

collecting and analyzing new soil cores is expensive (Rogge *et al.*, 2018). Apart from indicating zones with critically low SOC contents, SOC maps are particularly valuable for assessing the SOC variability in support of sample designs that are both representative for the region and minimize the number of sample points required (Conant and Paustian, 2002; de Gruijter *et al.*, 2016).

Xiao *et al.* (2019) provided an extensive overview of the role of remote sensing in measuring fluxes and stocks in the terrestrial C cycle. For quantifying SOC stocks, they describe two main techniques: (i) digital soil mapping (DSM) relying on empirical relationships between measured soil properties and spatially distributed co-variates (McBratney *et al.*, 2003) and (ii) imaging spectrometry of bare topsoils based on chemometric techniques. DSM techniques use legacy soil data and relate them to spatially distributed layers describing the so-called SCORPAN (soil, climate, organisms, relief, parent material, age and site) co-variates. Among these SCORPAN co-variates, organisms play a dominant role in explaining the variability in SOC at the regional to landscape scale (Lamichhane, Kumar and Wilson, 2019). Satellite remote sensing products are particularly effective in providing co-variates characterizing the role of organisms such as land use maps and indicators for C input from the vegetation (e.g. NDVI; Poggio *et al.*, 2013).

This chapter focuses on the direct remote sensing of SOC (i.e. optical remote sensing using hyperspectral and high-resolution multispectral sensors), developed since the early 2000s. For croplands in seedbed condition, the uppermost 20-30 cm of the soil is homogenized by annual tillage, and therefore, the reflectance of the soil surface represents the soil properties throughout the plow layer. The spectral characteristics of soils in the visible, near infrared and shortwave infrared domain (i.e. VNIR-SWIR: 400-2500 nm) have proven successful in predicting soil properties since the 1990s (e.g. Ben-Dor and Banin, 1995). Both physical and chemical properties are associated with the continuum spectral signature and specific absorption bands (Stenberg *et al.*, 2010). Soil organic carbon is a so-called chromophore related to soil color and displays absorption features in the visible spectrum (400-700 nm), while weak overtones and combinations of these vibrations are due to stretching and bending of covalent bonds in the NIR and SWIR (700-2500 nm) regions (Angelopoulou *et al.*, 2019). Soil is a complex system with a large variability in chemical and physical composition. The correlation between the soil chromophores and the chemical or physical properties of the soil is not straightforward. The chemometric approach is an empirical method with a strong spectroscopic basis and for which selected bands in the model have specific assignments (Chabrillat *et al.*, 2019). Data mining techniques are used to define these empirical relations between the reflectance at several wavelengths and the soil property. They consist of multivariate analysis of the spectra against the chemical/physical data through principal component analysis regression (Chang *et al.*, 2001), partial least squares regression (PLSR) (Zhao *et al.*, 2015), artificial neural networks (ANN) and random forest among other techniques. Since these techniques are empirical and derived from multivariate statistics, they apply to the population of samples for which they were developed. The accuracy of the models is assessed through (cross) validation using samples for which soil properties were analyzed by means of traditional laboratory techniques. The figures of merit that are most frequently reported are the $R^2$ of the predicted versus observed SOC contents, the root mean square error (RMSE), the ratio of performance to deviation (RPD, i.e. the standard deviation divided by the RMSE) and the ratio of performance to interquartile range (RPIQ, i.e. similar to the RPD, but using the interquartile range instead of the standard deviation (Bellon-Maurel *et al.*, 2010)). The quality of SOC maps, based on the SOC prediction using the spectra acquired in each grid-cell, is evaluated based on the model uncertainty. After all, the regression models have an inherent distribution of their results and for each pixel this distribution results in an uncertainty that can be mapped. Few examples of uncertainty in spectral mapping of soil properties exist (e.g. Gomez, Drost and Roger, 2015).

A data set containing soil properties analyzed in the laboratory together with their reflectance spectra of dry, ground samples acquired in stable laboratory conditions is referred to as a soil spectral library and is used to derive these multivariate prediction models. While such libraries were in the past often restricted to a limited study area, large-scale spectral libraries are becoming available covering entire countries such as Australia (Baldock *et al.*, 2013), USA (Wijewardane *et al.*, 2016) or Brazil (Demattê *et al.*, 2019), continents such as Europe (land use and coverage area frame survey: LUCAS; Tóth *et al.*, 2013) or Africa (Shepherd and Walsh, 2007) and even a global spectral library with voluntary contributions from many countries (Viscarra Rossel *et al.*, 2016). As the relation between chromophores and soil properties is not always straightforward, applying empirical models developed from a spectral library to predict SOC content for a sample collected anywhere within a large area with heterogeneous soils does not always give satisfactory results. For large regions with a range of soil types, the variation in one or more soil properties (e.g. iron hydroxides or sand content) can influence a prediction model for another soil property such as SOC. In order to deal with the heterogeneity of the soils, Nocita *et al.* (2014) used a limited number of spectra from the LUCAS library based on their similarity with the spectrum of the sample to be predicted and developed a so-called local PLSR model.

Recently, miniature spectrometers and multi-camera systems that can be mounted on unmanned aerial systems (UAS) became available. These systems are generally limited to the VNIR domain (400–1000 nm). Nevertheless, Aldana-Jague *et al.* (2016) were able to map the SOC content in a long-term trial in

Rothamsted (UK) using multispectral imagery (wavelengths 480–550–670–780–880–1000 nm). They obtained a $R^2$ of 0.95 and an RMSE of 0.31% during cross validation. Furthermore, satellite imagery is now freely available at relatively high spectral (nine bands) and spatial (20 m) resolution from the Sentinel-2 multispectral instrument (MSI) with a revisit time of five days. Hyperspectral satellites such as PRISMA (Loizzo et al., 2018) have been launched or are planned for the near future EnMAP (Guanter et al., 2015) or Shalom (Feingersh and Ben Dor, 2015). These recent developments mean that (hyperspectral) remote sensing imagery is now much more readily available as the flight campaigns are either flexible and low-cost such as for the UAS-borne systems or the images are freely available at regular intervals such as for the new generation of satellites. The potential for either routine mapping of a large number of fields (by UAS systems) or the mapping of large areas (by satellites) means that there is a need for automated procedures for the calibration of the spectral signal and dealing with disturbance of the surface conditions (vegetation, residues, roughness, moisture; Diek et al., 2019)

This chapter first reviews recent pilot studies covering limited areas often with exposed bare soils. Then we focus on the challenges for large-scale application of spectral mapping when the soil and parent material are heterogeneous and surface conditions are unknown. In order to deal with these constraints we discuss (i) calibration of spectral models based on large spectral libraries, (ii) surface conditions that disturb the soil signal and (iii) time series of images in order to delimit cropland fields and increase the extent of bare soil that can be mapped. Finally, a case study deals with a SOC prediction map derived from the spectra of a Sentinel-2 image and calibrated using the LUCAS spectral library.

## 2 Pilot studies of spectral SOC mapping

Pilot studies have demonstrated that the spectral signal of the soil surface as captured by the sensor can be calibrated using the soil properties of the corresponding pixels. For the calibration, composite samples are generally collected from the plow layer (0–15 cm) covering an area from 6 by 6 m (Gholizadeh et al., 2018), 1 m$^2$ (Guo et al., 2019) or 3 m radius (Castaldi et al., 2018) in order to take the uncertainty of geo-referencing into account. These samples are then analyzed in the laboratory using standard chemical or physical methodology. Such pilot studies assume that the surface conditions are homogeneous and that the reflectance can be related to one or more soil properties. Chabrillat et al. (2019) discussed the potential and limitations of imaging spectroscopy for mapping and monitoring soil properties, while Angelopoulou et al. (2019) reviewed pilot studies from remote sensing platforms ranging from UAS to airborne and satellite (Table 1). A tractor-driven

**Table 1** Summary of remote sensing platforms for SOC monitoring in terms of their benefits and drawbacks

| Platform | Benefits | Drawbacks |
|---|---|---|
| Satellites | • Obtain topsoil information from large areas<br>• Provide information for inaccessible areas<br>• Provide auxiliary data<br>• Consistent temporal resolution for creation of time series<br>• Short revisit time<br>• Provide free data | • Atmospheric absorptions interfering with the spectral measurements<br>• Low signal-to-noise ratio due to a short integration time over the target area<br>• Mixed pixels contain more than bare soil surface (e.g. vegetation)<br>• Need for geometric and atmospheric corrections |
| Airborne | • Provide information for inaccessible areas<br>• Few imagery instruments but becoming more available in the range of (1000–2500 nm )<br>• High spatial resolution | • Need for certain meteorological conditions for remote sensing applications<br>• Limitation of measurements only in a thin layer of topsoil<br>• Legal constraints for the flights<br>• High operational complexity<br>• High cost |
| LASs | • Flight plan can be scheduled according to weather condition<br>• High spatial resolution | • Limited flight duration<br>• Limited pay load<br>• Need for atmospheric, geometric corrections<br>• Legal constraints for the flights |

*Source*: Angelopoulou *et al.*, 2019.

on-line system of proximal sensing, not mentioned in this review, uses VNIR-SWIR sensors having a sapphire glass window and internal light source mounted at the bottom of a subsoiler. They are thus dragged through the topsoil and collect an on-the-go signal of the soil in the plow layer (Nawar and Mouazen, 2019). Airborne sensors were introduced in the 1990s (Ben-Dor and Banin, 1995) and collect data cubes for pixels of c. 2*2 m containing 200–300 spectral bands in the VNIR-SWIR of the soil surface. While proximal sensing techniques are used for within-field variability, airborne campaigns cover flight strips of up to 10–70 km including a range of soil types (e.g. Stevens *et al.* (2010); Guo *et al.* (2019)).

The Swiss-Belgian Airborne Prism Experiment (APEX) sensor has been flown in several campaigns across Europe (Schaepman *et al.*, 2015). Shi *et al.* (2020) calibrated a PLSR model to predict SOC using the APEX spectra, acquired on 2 September 2018, and the SOC contents as analyzed in the corresponding pixels in a 7 km-wide and 66 km-long SW-NE strip in the Belgian loam belt ($R^2$= 0.56; RMSE = 0.30 %; n = 43 for an independent validation; Fig. 1). They predicted the SOC contents only for the pixels with an NDVI between 0.10 and

**Figure 1** Zoom of the SOC map for the flight strip of the APEX campaign in central Belgium, September 2018. UTM coordinates in the 31N zone are given. Modified after Shi *et al.* (2020).

0.22. These pixels correspond mainly to bare cropland fields that have been plowed and seeded with winter cereals (Fig. 2). The remaining area is either under another land use (grassland, forest or urban) covered by summer crops such as sugar beet, maize and potato, or outside the NDVI range, probably as a result of emerging winter cereals. The zooms show both within-field patterns and differences between adjacent fields that provide a detailed insight in the SOC distribution of the plow layer. The spots with higher SOC content within the fields are traces of historic charcoal kilns dating from before the early nineteenth century when large parts of the areas were under production forest and charcoal was used in iron smelters (Hardy *et al.*, 2017). The clear difference in SOC contents between fields in the southwestern corner of the image are probably due to differences in historic management of the fields. In the same fields a tongue of higher SOC contents appears that is probably related to a third-order valley where the soil is relatively moist. Castaldi *et al.* (2018) provided a more in-depth discussion of similar patterns in SOC contents for fields in central Belgium using a digital terrain model and a semivariogram analysis.

The pilot study of Guo *et al.* (2019) covers a freshly plowed field of c 1.5 by 1 km in Iowa (USA). They acquired spectra from 50 soil samples (0–15 cm) on a

**Figure 2** Example of a cropland field in the APEX flight strip for which the SOC content has been predicted.

regular grid both in the laboratory and from an airborne Headwall sensor (400–1700 nm; SpecTIR LLC). They used the PLSR regression to first directly and later indirectly (i.e. by multiplying the predicted soil organic matter (SOM) content with the predicted bulk density) predict the SOC stock. Not surprisingly, the cross validation of the laboratory models yielded the best results. Overall, the $R^2$ of the prediction of the SOM contents was higher (0.75 in the laboratory and 0.54 airborne) than either the bulk density or the directly predicted SOC stocks ($R^2 = 0.66$ in the laboratory and 0.42 airborne). The indirect prediction of the SOC stocks gave similar performances. The authors attributed the satisfactory prediction of SOC stocks, which is rare in the literature, to the chromophores of SOM and the strong correlation between SOM and bulk density.

Gholizadeh *et al.* (2018) compared the performance of spectral prediction models acquired in the laboratory, from an airborne campaign using the CASI (370–1040 nm)-SASI (960–2440 nm) sensor (Itres Ltd., Calgary, Canada) and from a Sentinel-2 image. They studied four agricultural sites of c. 2 by

2 km throughout the Czech Republic. Before the airborne campaign and the Sentinel-2 overflight the fields were plowed and harrowed and devoid of vegetation or residues. The prediction accuracy based on lab spectroscopy, airborne and Sentinel-2 spectra was adequate for SOC (for Sentinel 2: RMSE for the independent validation set 0.08–0.23% and an RPD of 1.60–1.92). The best correlation for SOC was obtained with the Sentinel-2 bands B4 (665 nm, width 39nm), B5 (704 nm, width 20 nm), B11 (1610 nm, width 141 nm) and B12 (2186 nm, width 238 nm). The SOC maps produced by Sentinel 2 gave a more accurate prediction of the spatial patterns in zones with high SOC contents compared to the maps produced from the airborne sensor.

Based on an extensive literature review, Angelopoulou *et al.* (2019) summarized the main benefits and drawbacks of remote sensing platforms (Table 1). Up to now, airborne sensors combine the best spectra with high spatial resolution, but the complexity of organizing a flight campaign and its high costs are the main reasons that these images are only available for a few areas. Because of the constraints for planning the flight campaigns in advance, it becomes already more difficult to ensure that the atmospheric conditions are optimal during a period when croplands have been harvested and re-seeded. During this period the surface spectra are most likely to reflect the soil properties such as SOC concentration and the signal is not influenced by (partial) vegetation or residue cover, variation in roughness or soil moisture content.

## 3 Challenges for SOC mapping over large extents

### 3.1 Calibration using soil spectral libraries

Until now, calibration of remote sensing hyperspectral imagery relied on developing spectral transfer functions between the signal of the sensor and the soil property analyzed on samples of the corresponding pixel (see previous section and, for example, Gomez, Lagacherie and Coulouma, 2008; Guo *et al.*, 2019). However, as the remote sensing platforms cover larger areas (new generation of hyperspectral satellites) or become more flexible for the acquisition of fields with no prior information (such as UASs), dedicated sampling campaigns and analysis of the soil properties become the main constraints in cost and effort for producing the SOC maps. Hence, attempts have been made to transfer a calibration model based on laboratory spectra contained in one of the large soil spectral libraries (e.g. LUCAS) to the spectra acquired by the airborne sensor. As is the case for wet chemistry analyses, the challenge of the transfer of spectral models developed using a particular spectrometer to another was already recognized by Shenk and Westerhaus (1991). For transfer between laboratory spectra, a series of soil standards that are scanned by both the so-called master and slave instruments are used (Kopačková and Ben-Dor,

2016). The transfer of spectral models derived from spectral libraries to the signal of remote sensing instruments is more complex. After all, for remote sensing instruments, the reflectance is not directly measured, and instead it is calculated after reduction of the optical characteristics of the atmosphere. Parts of the spectra cannot be used from remote sensing instruments as the signal is absorbed by the water vapor in the atmosphere. The spectral resolution and the signal-to-noise ratio (the latter in particular in the SWIR region) are, in general, poorer for the remote sensing instruments. Even when the soil surface is dry, smooth and free from residues, its signal can still be different from the signal of a dry, sieved ground sample measured under controlled conditions in the laboratory. Below are three examples that illustrate different approaches for the transfer of models built using spectral libraries to the remote sensing instruments, airborne and UAS-borne sensors in this case.

Nouri et al. (2017) used several transfer functions whereby they measured both the spectra from a number of samples (so-called standards) in the laboratory and the spectra at the corresponding pixel acquired from the airborne sensor. Good predictions of clay content were obtained using different techniques, such as model updating Rep file, transfer by orthogonal projection and piecewise direct standardization, with only 15 standards for an area of c. 24 km². However, the transfer functions are specific to the flight campaign and have to be established for each future campaign.

Castaldi et al. (2018) observed that airborne spectra acquired over tarmac or white paved surface differed between flight campaigns (APEX in the Belgian loam belt in 2013 and 2015), even though simultaneously spectra acquired with an ASD spectrometer remained stable. Hence, they proposed a different approach, the so-called bottom up approach, in order to integrate the LUCAS spectral library for the calibration of the airborne APEX spectra. First, they used PLSR models to predict the SOC content in 54 calibration samples based on the spectra of the LUCAS library, and then they used these predicted SOC contents to develop a calibration model using the APEX spectra of the corresponding pixels. This approach yielded SOC maps for the flight strips in Luxembourg and the Belgian loam belt with a satisfactory level of accuracy (RMSE = 1.5-4.9 g kg⁻¹; RPD = 1.4-1.7).

Although the spectral range of most UAS-borne spectrometers or multispectral cameras is limited to the VNIR region (400-1000 nm), there are certain advantages compared to the airborne sensors (Table 1). Before take-off, the spectrometer can be calibrated using a white reference (spectralon) meaning that the reflectance is measured and not calculated as is the case for airborne instruments. Moreover, a soil standard can be scanned before each flight. Kopačková and Ben-Dor (2016) suggest using perfect silica sand from an Australian beach as soil standard. Crucil et al. (2019) used the soil standard both in outdoor scans with a STS-VIS (350-800 nm; Ocean Optics Inc., Dunedin,

FL, USA) and in the laboratory using an ASD Fieldspec 3 FR spectro-radiometer (Analytical Spectral Devices Inc.). They demonstrated that using a local spectral library for a long-term trial in Rothamsted (UK) the soil standard allowed applying the laboratory model to predict the SOC content of the long-term trial in outdoor conditions using the small STS spectrometer. Further tests with the STS spectrometer flown by a UAS are ongoing.

The examples in the previous paragraphs dealt with aligning the spectra acquired by remote sensing instruments to the ones of a spectral library. The heterogeneity of the soil and parent material determines the selection of optimal pattern and number of samples for the calibration of spectral models. Optimal sampling schemes aim to cover the variation in spectral characteristics of the study area (i.e. the feature space) rather than covering its spatial extent (i.e. the geographical space; Ramirez-Lopez et al. (2014)). Castaldi, Chabrillat and van Wesemael (2019) demonstrated the use of Sentinel-2 imagery in order to characterize the spectral feature space of a target area. They advocate using algorithms that simultaneously calculate the number of samples and their spatial distribution such as the Puchwein algorithm. Thus, an efficiency of sample selection, expressed as ratio between accuracy and sampling density can be calculated.

When large spectral libraries, such as LUCAS topsoil database, exist, the heterogeneity of the soils and parent materials prevent a straightforward interpretation of the spectra in terms of soil properties. The application of these spectral libraries for predicting SOC in a specific cropland site requires a selection of the spectra that best reflect those of the target area. Ward et al. (2019) applied both a k-means clustering and a local PLSR approach in order to build spectral models that do not rely on external co-variates. Ward et al. (2019) removed the water vapor bands from the LUCAS spectra and re-sampled them to the spectral resolution of the forthcoming EnMAP satellite in order to evaluate the potential for using the LUCAS spectral library to build calibration models for EnMAP. Overall, the local PLSR re-sampled to the EnMAP characteristics gave good results ($R^2$=0.66, RMSEP = 5.78 g kg$^{-1}$ and RPIQ = 1.93).

### 3.2 Dealing with surface conditions that disturb the soil signal

For most of the pilot studies it was assumed that the pixels on the SOC map represented smooth, bare and dry soils. For these pilot studies, cropland soils were targeted when they had just been harrowed and seeded (Chabrillat et al., 2019). However, the availability of and the area covered by the images provided by satellite systems such as MSI/Sentinel-2 and the upcoming hyperspectral satellites has increased dramatically. Vaudour et al. (2019), working in the Versailles plain (France) demonstrated the impact of acquisition date of Sentinel-2 imagery and related weather and surface conditions on the prediction of topsoil SOC in croplands. Not surprisingly, they derived the best SOC models for spring images when the soil moisture and soil roughness were lowest. NDVI values did not influence the prediction performance as long as they were below a threshold of 0.35. This example demonstrates that there is a need for automated procedures that ensure that the pixels under consideration represent dry, smooth and bare cropland soils (Diek et al., 2019). After all, only soils in these conditions are similar to the dry, ground samples scanned to populate the spectral libraries. At this stage, it will most likely not be possible (or only to a limited extent) to correct for roughness, moisture, green and dry vegetation or residues and therefore we suggest to mask all pixels that do not represent similar conditions to the ones in the spectral library (Chabrillat et al., 2019). Here we will only discuss the methods to evaluate soil surface conditions using spectral indices.

The spectral signal of the earth surface is anisotropic. Roughness produces shadow areas where solar beams do not reach the surface, and hence the reflectance is much lower (Chabrillat et al., 2019). It is particularly difficult to correct for the shading effects in furrows created by tillage, as their orientation toward the sun and the sensor determines the degree of shading and reduction of the reflectance. Although techniques exist for creating detailed digital terrain models representing the furrows, as far as we know, no attempts have been made to use these DTMs for each individual field in order to correct the tillage-induced reflection. For temperate croplands, summer crops are sown in April and winter crops are sown in September (see, for example, Diek et al., (2016)). Depending on the weather conditions these are the months with the largest extent of bare soils. Nowadays, farmers harrow and seed their fields within hours after plowing, and therefore most bare soils are smooth during the harvest and seeding periods.

Soil moisture has a profound effect on the shape of a spectrum, as it not only reduces the overall albedo, but also enhances the water absorption peaks around 1400 nm and 1920 nm. Minasny et al. (2011) developed the external parameter orthogonalization approach (EPO) in order to split a spectrum of a moist soil in a spectrum that can be attributed to its inherent soil properties and one that reflects the effect of its moisture content. Unfortunately, the EPO approach cannot be used for assessing the variability in soil moisture content of remote sensing image. First of all, the atmospheric moisture produces so much noise at the wavelengths of the moisture absorption peaks that they have to be deleted prior to further treatment of the spectra. Then, the EPO approach requires spectra, acquired with the same instrument, from both dry soils and soils with varying soil moisture content. When using remote sensing instruments the only spectra that can be acquired are the ones of the soil surface with an unknown moisture content. Therefore, Haubrock et al. (2008) suggested characterizing the soil moisture content through the normalized soil moisture index (NSMI) using the shoulders of the water absorption peak in the SWIR (eq. 1).

$$NSMI = \frac{Ri - Rj}{Ri + Rj} \tag{1}$$

Where $R_i$ and $R_j$ are the reflectance values at 1800 nm and 2119 nm.

A linear relation between NSMI and gravimetric soil moisture content ($R^2$ = 0.7; RMSE 0.03 g g$^{-1}$) was obtained by Haubrock *et al.* (2008) and Nocita *et al.* (2013). Nowadays, soil surface moisture can be cautiously estimated based on the NSMI. Validation of NSMI maps derived from remote sensing instruments remains difficult, as the instruments only register the surface moisture content, while other techniques (e.g. capacitive soil moisture probes, ground penetrating radar, sampling and gravimetric analysis) consider a volume of soil that extends to at least 5-10 cm. During a sunny, cloud free day required for a useful satellite image the top 1-2 cm of a bare cropland can quickly dry out creating a steep soil moisture gradient. Hence, correcting for the variation in soil moisture when developing prediction models for organic or mineral soil properties applied to entire remote sensing images is still not possible (Chabrillat *et al.*, 2019). Nevertheless, several authors used laboratory and field instruments measuring the spectra from soils with a range of moisture contents to demonstrate that SOC prediction improves when separate models are developed by NSMI class (Nocita *et al.*, 2013; Rodionov *et al.*, 2014; Fabre *et al.*, 2015; Bablet *et al.*, 2018; Yue *et al.*, 2019).

SOC prediction models are very sensitive to fractional cover of the soil surface by green vegetation, dry vegetation or residues and soil surface crusts (Bartholomeus *et al.*, 2011; Rodionov *et al.*, 2014; Chabrillat *et al.*, 2019). In particular, green, so-called photosynthetically active, vegetation, has a large influence on the performance of SOC models. Bartholomeus *et al.* (2011) reported a relative error of 58% in SOC prediction for a cover of 5-10%, while Rodionov *et al.* (2014) found that SOC was over estimated by 30 g kg$^{-1}$ for only 5% surface cover by green leaves. While Bartholomeus *et al.* (2011) proposed a spectral unmixing approach to correct SOC predictions for vegetation cover up to 25%, Rodionov *et al.* (2014) calculated correction values based on field spectra obtained in long-term trial with a range of vegetation cover. The NDVI effectively separates the signal of green vegetation from the soil signal using the sharp increase in reflectance at the red edge (at 700 nm Fig. 3). To characterize straw cover, they use the cellulose absorption index (CAI) based on the cellulose absorption at wavelengths around 2000 and 2400 nm (eq. 2; Fig. 3).

$$CAI = 0.5(R_{2.0} + R_{2.2}) - R_{2.1} \tag{2}$$

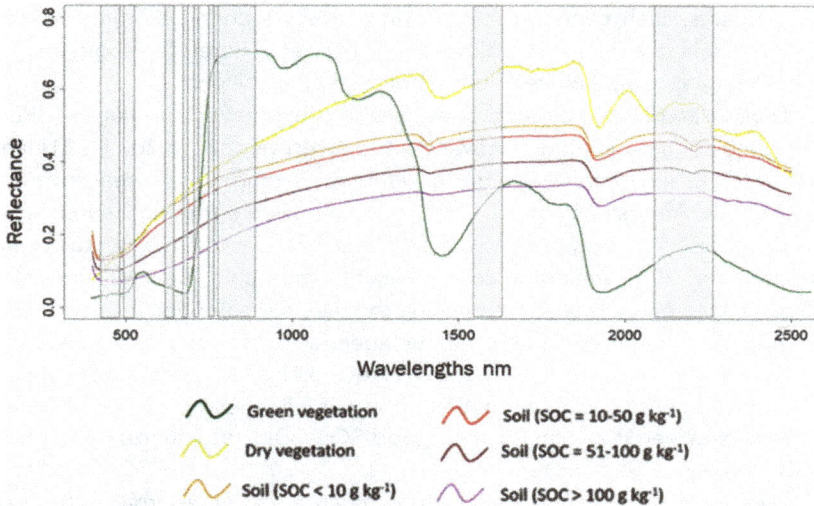

**Figure 3** Laboratory soil spectra (400–2500 nm) from LUCAS topsoil database with different organic carbon content compared to dry and green vegetation spectra. The vertical grey bars indicate the Sentinel-2 MSI bands (Castaldi et al., 2019).

Where $R_{2.0}$, $R_{2.2}$, $R_{2.1}$ are reflectance factors in bands at 2000–2050 nm, 2089–2130 nm and 2190–2240 nm, respectively (Rodionov et al., 2016). As a bare soil does not have an absorption peak at 2100 nm, its CAI is usually negative in contrast to the positive CAI of a straw-covered soil. Unfortunately, satellite sensors, such as Landsat or Sentinel-2, do not very well cover the SWIR wavelengths (Fig. 3). The normalized burn ratio 2 (NBR2) index based on the two SWIR bands uses wavelengths (e.g. for Sentinel-2: B11 (center 1610.4 nm, bandwidth 141 nm) and B12 (center 2185.7 nm, bandwidth 238 nm) are somewhat similar to the ones used for the NSMI and the CAI index (Fig. 3).

Ben-Dor et al. (2004) have shown that soil crusts and their implication for infiltration rates can be determined from hyperspectral imagery of loess agricultural soils. These crusts develop gradually on cropland soils as a result of raindrop impact throughout the growing season. Thus, well-developed crusts will occur on fields that have been harvested, but they are destroyed by tillage and the preparation of the seedbed. Soil crusts in agricultural soils mostly co-exist with stubble and residue cover, and therefore these conditions can be detected using the CAI (eq. 2).

# 4 Synthetic bare soil images

One of the main disadvantages of spectral mapping of soil properties is that the maps rely on the spectral signal of bare soils. In particular, in temperate environments, such conditions prevail in croplands during a limited period when the soils are harvested, plowed and seeded. Therefore, single images typically contain only a limited number of such bare soil pixels.

Diek *et al.* (2016) acquired three airborne hyperspectral images (APEX) during the spring and autumn sowing and harvest periods in 2013, 2014 and 2015 with the aim of creating a mosaic and thus extending the areal coverage of bare soils. After selecting the bare soil pixels using a mask based on NDVI and CAI thresholds, they applied the empirical line method in order to correct the 2013 and 2015 spectra in such a way that they coincided with the 2014 spectrum for a number of pixels meeting the bare soil condition in each image. The resulting image contains synthetic spectra for the bare soil pixels that cover twice the area compared to the one covered by a single image. However, the synthetic spectra were probably too much influenced by variation in soil moisture between the years to yield good SOC prediction models ($R^2 = 0.39$ for the three year composite).

Rogge *et al.* (2018) developed a multi-temporal approach, the soil composite mapping processor (SCMAP), based on 15 years of LandSat imagery for the definition of croplands. Croplands change from exposed soil (after plowing) to photosynthetically active vegetation cover at least once per year. For multi-temporal composites, they defined minimum and maximum NDVI thresholds to separate cropland from other land uses. They tested the approach in Germany and extend it to other countries. Surface reflectance composites can be used to evaluate management practices such as the frequency of temporary grassland in the rotation or the management of crop residues, while averages of reflectance of exposed soil pixels relate to broad soil groups.

Demattê *et al.* (2018) proposed the geospatial soil sensing system (GEOS3) in order to bring together the fragmented information of bare soils in Brazil into a single synthetic image. The procedure is based on the following steps: (i) collection of Landsat 5 legacy data, (ii) selection of images from the dry seasons, (iii) development of a set of rules based on the NDVI and NBR2 to select bare soils, (iv) calculation of a median spectrum based from the time series of images for each exposed soil pixel, (v) aggregation to a synthetic soil image and (vi) validation of the quality of this image through correlation between laboratory and median spectra of the exposed soil pixels. While only 5% of the area was classified as bare soil in a single image, this area increased to 68% of the synthetic image.

These techniques to create synthetic images from time series of satellite images clearly result in more continuous reflectance maps. These delineate the croplands, that is, the only areas with a strong annual variation in NDVI. Furthermore, the median reflectance of exposed soils in the time series already shows a correlation to the main soil types. Unfortunately, these median spectra are influenced by the variation in soil moisture between the individual images and cannot, yet, be used for SOC prediction models.

# 5 Case study

The case study is an example of high-resolution mapping of the SOC content in the topsoil of croplands (0–20 cm) covering a large area (100*100 km) at low costs (Castaldi, Chabrillat, Don *et al.*, 2019). A Sentinel-2B image was acquired on 30 August, 2017, covering the area around the town of Demmin in northern Germany, part of the Tereno NE long-term monitoring site (Zacharias *et al.*, 2011). This image was selected because it was largely cloud-free and the time window corresponded to a period in which the soils are likely to be in seedbed condition after sowing of winter cereals. The site is dominated by cropland (80.3%) and part of the young morainic soil-landscape of northern Germany characterized by geomorphological features, including (sandy-) loamy morainic parent material of the late Pleistocene, kettle holes and slightly undulating relief. The calibration data set consisted of samples in croplands covered by the Sentinel-2 image extracted from the LUCAS spectral library. The laboratory spectra were re-sampled according to the spectral bands of the MSI/Sentinel-2B sensor. They used a combination of two bands of the re-sampled spectra to fit an exponential regression model to the SOC content. These models were later applied to real MSI/Sentinel-2 spectra to predict the SOC content in the bare pixels of the satellite image. Bare pixels meet all of the following criteria: (i) zero clouds probability, (ii) NDVI lower than 0.35, (iii) differences between band 3 (B3) and B2 as well as between B4 and B3 larger than 0 in order to improve the soil mask and (iv) different NBR2 index thresholds were used to exclude spectra from pixels with high moisture content and covered by straw or dry vegetation. The validation data set consisted of a maximum of 253 samples collected in the area as part of the Tereno long-term monitoring and the German agricultural soil inventory (Vos *et al.*, 2019).

Overall, 12% of the pixels were covered by clouds, and of the cloud free area 26% of the pixels was not covered by green vegetation. The best ratio of performance to deviation (RPD) was obtained using the average between bands B6 (740 nm) and B5 (704 nm; RPD: 4.4; RMSE 6.8 g kg$^{-1}$) and between B4 (665 nm) and B5 (RPD: 2.9; RMSE 10.3 g kg$^{-1}$) adopting a very strict NBR2 threshold (0.05). The strict NBR2 threshold reduced the area that could be mapped to 10% of the pixels. Increasing the NBR2 to 0.1 resulted in a larger area that could be mapped (15%) at the expense of the accuracy (RMSE 15.7 g kg$^{-1}$). This application uses a spectral model derived from the LUCAS spectral library that is then applied to the MSI/Sentinel 2B sensor. In contrast to the pilot studies, there is no need for sampling and analysis of SOC content. In theory, the approach could be applied to other Sentinel-2 images within the area covered by the LUCAS spectral library. However, the soils should preferably be quite homogeneous in terms of parent material and pedogenic oxides in order to avoid the need for clustering or local regressions (Ward *et al.*, 2019). Moreover, as the NBR2 index is less specific than the CAI or NSMI indices, its capacity to distinguish between the effect of residues and soil moisture still has to be investigated. In any case Vaudour *et al.* (2019) have demonstrated that the performance of SOC predictions based on Sentinel-2 are poor under suboptimal conditions, that is, in winter, for freshly plowed soils.

# 6 Summary and future trends

Spectral mapping of soil properties and in particular SOC has provided high-resolution maps of bare croplands for pilot studies. Although strictly speaking only the signal of the soil surface is registered, most authors have worked on fields that have been harrowed and seeded. Thus, conditions that disturb the soil signal such as roughness, variation in moisture content, residues or vegetation hardly influence the performance of the prediction models for well-controlled pilot studies. As more imagery becomes available with the launch of hyperspectral and high resolution multispectral satellites and the miniaturization of spectrometers that can be mounted on UASs, there is a need for automated procedures for calibration and for detection of conditions disturbing the soil signal. Spectral models are already calibrated using spectral libraries (e.g. LUCAS) and clustering or local PLSR approaches have been developed to deal with the heterogeneity of the soil matrix that in the past often led to inaccurate prediction of the soil property. Transfer of these models developed under laboratory conditions to the signal of bare soil pixel acquired by a remote sensor is still in a research stage. Spectral indices that allow distinguishing a bare, dry soil from a wet soil or a soil covered with green vegetation or residue exist and are mainly based on the NDVI and a number of indices using SWIR bands.

The Sentinel 2 program is a unique opportunity for the development of SOC prediction models. Pilot studies have demonstrated that adequate SOC models can be derived for agricultural areas when the soils have been plowed and harrowed. The Czech case study demonstrated that due to its high spatial resolution and stable signal, the SOC maps produced with a spectral model based on Sentinel-2 were even better than the ones using spectra from an airborne platform. The short revisit time (5 days) increases the potential for the acquisition of (a series of) images covering a large extent of exposed soils. Although an adequate SOC prediction model for an entire Sentinel-2 image in Northern Germany was already produced, the possibility to use composite images consisting of pixels acquired at different dates and thus extending the spatial coverage still needs to be confirmed. In order to facilitate the processing of Sentinel 2 (and future hyperspectral) satellite imagery at the regional scale, current research focuses on the transfer of spectral models calibrated on large-scale spectral libraries to the signal of remote sensing instruments. This is vital as for routine applications a large-scale spectral library would reduce (or even eliminate) the cost for sampling and analysis and would provide a standardization of the calibration, as the data in the spectral library are all acquired using a uniform protocol. Furthermore, the focus is on the development of automated procedures to distinguish bare cropland soil pixels, as these are the ones to which the spectral SOC models can be applied. The NDVI can be easily calculated from the spectra of multispectral satellites. However, the indices that characterize soil moisture or residue cover use SWIR bands that are not registered by the current generation of satellites. Last but not least, the conditions for acquiring imagery that best reflect the composition of the bare soil have not yet been objectively defined, although the analysis of time series has provided cropland masks and synthetic images of multiple dates and tending towards a continuous cropland cover.

## 7 Where to look for further information

### 7.1 Further reading

A comprehensive review of the latest research on remote sensing for SOC estimation is given by Angelopoulou et al. (2019). For full bibliographic details we refer to the next section (References).

An overview of the principles behind imaging spectroscopy for soil mapping and monitoring is given by Chabrillat et al. (2019).

Another overview of remote sensing applied to soils is given by Dematte et al. (2015) in Thenkabail, P. S. (ed.). *Remote sensing handbook–three volume set: land resources monitoring, modeling, and mapping with remote sensing*. CRC Press, Boca Raton, pp. 661–732. ISBN-10: 1482217953.

### 7.2 Key conferences

The European Association of Remote Sensing Laboratories (EARSel) has a special interest group on imaging spectroscopy (SIG IS; http://atcor.dlr.de/SIG -IS.htm). This group organizes bi-annual conferences on hyperspectral remote sensing for mapping and monitoring soil properties.

### 7.3 Useful websites

Sentinel-2 imagery is freely available and can be accessed through the Copernicus open access hub (https://scihub.copernicus.eu/).

The Sentinel-2 Toolbox can be used for the visualization and analysis of products. More information can be found at: https://sentinel.esa.int/web/s entinel/toolboxes/sentinel-2.

Information on freely available Landsat images can be found at https://la ndsat.gsfc.nasa.gov/data/where-to-get-data/.

The German Aerospace Center (DLR) has developed a tool for mapping exposed soil surfaces based on Landsat imagery referred to as the soil composite processor (ScMaP): https://geoservice.dlr.de/data-assets/c5ng9yn nyx29.html.

The European soil spectral library developed within the LUCAS soil framework is provided by the EU Joint Research Center. Please consult the website from the European Soil Data Center (ESDAC): https://esdac.jrc.ec .europa.eu/.

The Hyperspectral SOil Mapper (HYSOMA) software interface is developed by the German Research Center for Geosciences (GFZ) and can be downloaded from http://www-app2.gfz-potsdam.de/hysoma.

Antoine Stevens and Leonardo Ramirez Lopez (2014) wrote a useful R package for pre-treating spectra and selecting calibration algorithms: Prospectr at https://cran.r-project.org/web/packages/prospectr/vignettes/prospectr-intro.pdf.

# 8 References

Aldana-Jague, E., Sommer, M., Saby, N. P. A., Cornelis, J.-T., van Wesemael, B. and van Oost, K. (2016). High resolution characterization of the soil organic carbon depth profile in a soil landscape affected by erosion. *Soil and Tillage Research* 156, 185–193. https://doi.org/10.1016/j.still.2015.05.014.

Angelopoulou, T., Tziolas, N., Balafoutis, A., Zalidis, G. and Bochtis, D. (2019). Remote sensing techniques for soil organic carbon estimation: a review. *Remote Sensing* 11(6). https://doi.org/10.3390/rs11060676.

Bablet, A., Vu, P. V. H., Jacquemoud, S., Viallefont-Robinet, F., Fabre, S., Briottet, X., Sadeghi, M., Whiting, M. L., Baret, F. and Tian, J. (2018). MARMIT: a multilayer radiative transfer model of soil reflectance to estimate surface soil moisture content in the solar domain (400–2500 nm). *Remote Sensing of Environment* 217, 1–17. https ://doi.org/10.1016/j.rse.2018.07.031.

Baldock, J. A., Hawke, B., Sanderman, J. and MacDonald, L. M. (2013). Predicting contents of carbon and its component fractions in Australian soils from diffuse reflectance mid-infrared spectra. *Soil Research* 51(7–8), 577–595. https://doi.org/10.1071/ SR13077.

Bartholomeus, H., Kooistra, L., Stevens, A., van Leeuwen, M., van Wesemael, B., Ben-Dor, E. and Tychon, B. (2011). Soil organic carbon mapping of partially vegetated agricultural fields with imaging spectroscopy. *International Journal of Applied Earth Observation and Geoinformation* 13(1), 81–88. https://doi.org/10.1016/j.jag.2010. 06.009.

Bellon-Maurel, V., Fernandez-Ahumada, E., Palagos, B., Roger, J.-M. and McBratney, A. (2010). Critical review of chemometric indicators commonly used for assessing the quality of the prediction of soil attributes by NIR spectroscopy. *TrAC Trends in Analytical Chemistry* 29(9), 1073–1081. https://doi.org/10.1016/j.trac.2010.05.006.

Ben-Dor, E. and Banin, A. (1995). Near infrared analysis (NIRA) as a method to simultaneously evaluate spectral featureless constituents in soils. *Soil Science* 159(4), 259–270. https://doi.org/10.1097/00010694-199504000-00005.

Ben-Dor, E., Goldshleger, N., Braun, O., Kindel, B., Goetz, A. F. H., Bonfil, D., Margalit, N., Binaymini, Y., Karnieli, A. and Agassi, M. (2004). Monitoring infiltration rates in semiarid soils using airborne hyperspectral technology. *International Journal of Remote Sensing* 25(13), 2607–2624. https://doi.org/10.1080/0143116031000164 2322.

Castaldi, F., Chabrillat, S., Don, A. and van Wesemael, B. (2019). Soil organic carbon mapping using LUCAS topsoil database and Sentinel-2 data: an approach to reduce soil moisture and crop residue effects. *Remote Sensing* 11(18). https://doi.org/10 .3390/rs11182121.

Castaldi, F., Chabrillat, S., Jones, A., Vreys, K., Bomans, B. and van Wesemael, B. (2018). Soil organic carbon estimation in croplands by hyperspectral remote APEX data using the LUCAS topsoil database. *Remote Sensing* 10(2). https://doi.org/10.3390 /rs10020153.

Castaldi, F., Chabrillat, S. and van Wesemael, B. (2019). Sampling strategies for soil property mapping using multispectral Sentinel-2 and hyperspectral EnMAP satellite data. *Remote Sensing* 11(3). https://doi.org/10.3390/rs11030309.

Chabrillat, S., Ben-Dor, E., Cierniewski, J., Gomez, C., Schmid, T. and van Wesemael, B. (2019). Imaging spectroscopy for soil mapping and monitoring. *Surveys in Geophysics* 40(3), 361–399. https://doi.org/10.1007/s10712-019-09524-0.

Chang, C.-W., Laird, D. A., Mausbach, M. J. and Hurburgh, C. R., Jr. (2001). Near-infrared reflectance spectroscopy - principal components regression analyses of soil properties. *Soil Science Society of America Journal* 65(2), 480–490.

Chenu, C., Angers, D. A., Barré, P., Derrien, D., Arrouays, D. and Balesdent, J. (2019). Increasing organic stocks in agricultural soils: knowledge gaps and potential innovations. *Soil and Tillage Research* 188, 41–52. https://doi.org/10.1016/j.still.201 8.04.011.

Conant, R. T. and Paustian, K. (2002). Spatial variability of soil organic carbon in grasslands: implications for detecting change at different scales. *Environmental Pollution* 116 (Suppl. 1), S127–S135. https://doi.org/10.1016/S0269-7491(01)00265-2.

Crucil, G., Castaldi, F., Aldana-Jague, E., van Wesemael, B., Macdonald, A. and Oos, K. V. (2019). Assessing the performance of UAS-Compatible multispectral and hyperspectral sensors for soil organic carbon prediction. *Sustainability* 11(7). https:// doi.org/10.3390/su11071889.

de Gruijter, J. J., McBratney, A. B., Minasny, B., Wheeler, I., Malone, B. P. and Stockmann, U. (2016). Farm-scale soil carbon auditing. *Geoderma* 265, 120–130. https://doi.org /10.1016/j.geoderma.2015.11.010.

Demattê, J. A. M., Dotto, A. C., Paiva, A. F. S., Sato, M. V., Dalmolin, R. S. D., de Araújo, MdS. B., da Silva, E. B., Nanni, M. R., ten Caten, A., Noronha, N. C., Lacerda, M. P. C., de Araújo Filho, J. C., Rizzo, R., Bellinaso, H., Francelino, M. R., Schaefer, C. E. G. R., Vicente, L. E., dos Santos, U. J., de Sá Barretto Sampaio, E. V., Menezes, R. S. C., de Souza, J. J. L. L., Abrahão, W. A. P., Coelho, R. M., Grego, C. R., Lani, J. L., Fernandes, A. R., Gonçalves, D. A. M., Silva, S. H. G., de Menezes, M. D., Curi, N., Couto, E. G., dos Anjos, L. H. C., Ceddia, M. B., Pinheiro, ÉF. M., Grunwald, S., Vasques, G. M.,

Marques Júnior, J., da Silva, A. J., Barreto, M. CdV., Nóbrega, G. N., da Silva, M. Z., de Souza, S. F., Valladares, G. S., Viana, J. H. M., da Silva Terra, F., Horák-Terra, I., Fiorio, P. R., da Silva, R. C., Frade Júnior, E. F., Lima, R. H. C., Alba, J. M. F., de Souza Junior, V. S., Brefin, M. D. L. M. S., Ruivo, M. D. L. P., Ferreira, T. O., Brait, M. A., Caetano, N. R., Bringhenti, I., de Sousa Mendes, W., Safanelli, J. L., Guimarães, C. C. B., Poppiel, R. R., e Souza, A. B., Quesada, C. A. and do Couto, H. T. Z. (2019). The Brazilian Soil Spectral Library (BSSL): a general view, application and challenges. *Geoderma* 354. https://doi.org/10.1016/j.geoderma.2019.05.043.

Demattê, J. A. M., Fongaro, C. T., Rizzo, R. and Safanelli, J. L. (2018). Geospatial Soil Sensing System (GEOS3): a powerful data mining procedure to retrieve soil spectral reflectance from satellite images. *Remote Sensing of Environment* 212, 161–175. https://doi.org/10.1016/J.RSE.2018.04.047.

Demattê, J. A. M., Morgan, C. L. S., Chabrillat, S., Rizzo, R., Franceschini, M. H. D., Terra, F. d., Vasques, G. M. and Wetterlind, J. (2015). Spectral sensing from ground to space in soil science: state of the art, applications, potential and perspectives. In: Thenkabail, P. S. (Ed.), *Remote Sensing Handbook–Three Volume Set: Land Resources Monitoring, Modeling, and Mapping with Remote Sensing*. CRC Press, Boca Raton, pp. 661–732. ISBN-10: 1482217953.

Diek, S., Chabrillat, S., Nocita, M., Schaepman, M. E. and de Jong, R. (2019). Minimizing soil moisture variations in multi-temporal airborne imaging spectrometer data for digital soil mapping. *Geoderma* 337, 607–621. https://doi.org/10.1016/j.geoderma.2018.09.052.

Diek, S., Schaepman, M. E. and de Jong, R. (2016). Creating multi-temporal composites of airborne imaging spectroscopy data in support of digital soil mapping. *Remote Sensing* 8(11). https://doi.org/10.3390/rs8110906.

Fabre, S., Briottet, X. and Lesaignoux, A. (2015). Estimation of Soil Moisture Content from the Spectral Reflectance of Bare Soils in the 0.4–2.5 µm Domain. *Sensors* 15(2), 3262–3281. https://doi.org/10.3390/s150203262.

Feingersh, T. and ben Dor, E. (2015). Shalom – a commercial hyperspectral space mission. In: *Optical Payloads for Space Missions* (pp. 247–263). John Wiley & Sons, Ltd. https://doi.org/10.1002/9781118945179.ch11.

Gholizadeh, A., Žižala, D., Saberioon, M. and Borůvka, L. (2018). Soil organic carbon and texture retrieving and mapping using proximal, airborne and Sentinel-2 spectral imaging. *Remote Sensing of Environment* 218, 89–103. https://doi.org/10.1016/j.rse.2018.09.015.

Gomez, C., Drost, A. P. A. and Roger, J. -M. (2015). Analysis of the uncertainties affecting predictions of clay contents from VNIR/SWIR hyperspectral data. *Remote Sensing of Environment* 156, 58–70. https://doi.org/10.1016/j.rse.2014.09.032.

Gomez, C., Lagacherie, P. and Coulouma, G. (2008). Continuum removal versus PLSR method for clay and calcium carbonate content estimation from laboratory and airborne hyperspectral measurements. *Geoderma* 148(2), 141–148. https://doi.org/10.1016/J.GEODERMA.2008.09.016.

Guanter, L., Kaufmann, H., Segl, K., Foerster, S., Rogass, C., Chabrillat, S., Kuester, T., Hollstein, A., Rossner, G., Chlebek, C., Straif, C., Fischer, S., Schrader, S., Storch, T., Heiden, U., Mueller, A., Bachmann, M., Mühle, H., Müller, R., Habermeyer, M., Ohndorf, A., Hill, J., Buddenbaum, H., Hostert, P., van der Linden, S., Leitão, P., Rabe, A., Doerffer, R., Krasemann, H., Xi, H., Mauser, W., Hank, T., Locherer, M., Rast, M., Staenz, K. and

Sang, B. (2015). The EnMAP spaceborne imaging spectroscopy mission for earth observation. *Remote Sensing* 7(7), 8830–8857. https://doi.org/10.3390/rs70708830.

Guo, L., Zhang, H., Shi, T., Chen, Y., Jiang, Q. and Linderman, M. (2019). Prediction of soil organic carbon stock by laboratory spectral data and airborne hyperspectral images. *Geoderma* 337, 32–41. https://doi.org/10.1016/J.GEODERMA.2018.09.003.

Hardy, B., Cornelis, J.-T., Houben, D., Leifeld, J., Lambert, R. and Dufey, J. E. (2017). Evaluation of the long-term effect of biochar on properties of temperate agricultural soil at pre-industrial charcoal kiln sites in Wallonia, Belgium. *European Journal of Soil Science* 68(1), 80–89. https://doi.org/10.1111/ejss.12395.

Haubrock, S.-N., Chabrillat, S., Lemmnitz, C. and Kaufmann, H. (2008). Surface soil moisture quantification models from reflectance data under field conditions. *International Journal of Remote Sensing* 29(1), 3–29. https://doi.org/10.1080/01431160701294695.

Kibblewhite, M. G., Ritz, K. and Swift, M. J. (2008). Soil health in agricultural systems. In: *Philosophical Transactions of the Royal Society B: Biological Sciences* 363(1492), 685–701. Royal Society. https://doi.org/10.1098/rstb.2007.2178.

Kopačková, V. and Ben-Dor, E. (2016). Normalizing reflectance from different spectrometers and protocols with an internal soil standard. *International Journal of Remote Sensing* 37(6), 1276–1290. https://doi.org/10.1080/01431161.2016.1148291.

Lamichhane, S., Kumar, L. and Wilson, B. (2019). Digital soil mapping algorithms and covariates for soil organic carbon mapping and their implications: a review. *Geoderma* 352, 395–413. https://doi.org/10.1016/J.GEODERMA.2019.05.031.

Loizzo, R., Guarini, R., Longo, F., Scopa, T., Formaro, R., Facchinetti, C. and Varacalli, G. (2018). Prisma: the Italian hyperspectral mission. *International Geoscience and Remote Sensing Symposium (IGARSS)*, 2018-July, 175–178. https://doi.org/10.1109/IGARSS.2018.8518512.

McBratney, A. B., Mendonça Santos, M. L. and Minasny, B. (2003). On digital soil mapping. *Geoderma* 117(1–2), 3–52. https://doi.org/10.1016/S0016-7061(03)00223-4.

Minasny, B., Malone, B. P., McBratney, A. B., Angers, D. A., Arrouays, D., Chambers, A., Chaplot, V., Chen, Z., Cheng, K., Das, B. S., Field, D. J., Gimona, A., Hedley, C. B., Hong, S. Y., Mandal, B., Marchant, B. P., Martin, M., McConkey, B. G., Mulder, V. L., O'Rourke, S., Richer-de-Forges, A. C., Odeh, I., Padarian, J., Paustian, K., Pan, G., Poggio, L., Savin, I., Stolbovoy, V., Stockmann, U., Sulaeman, Y., Tsui, C., Vågen, T., van Wesemael, B. and Winowiecki, L. (2017). Soil carbon 4 per mille. *Geoderma* 292, 59–86. https://doi.org/10.1016/j.geoderma.2017.01.002.

Minasny, B., McBratney, A. B., Bellon-Maurel, V., Roger, J.-M., Gobrecht, A., Ferrand, L. and Joalland, S. (2011). Removing the effect of soil moisture from NIR diffuse reflectance spectra for the prediction of soil organic carbon. *Geoderma* 167–168, 118–124. https://doi.org/10.1016/J.GEODERMA.2011.09.008.

Nawar, S. and Mouazen, A. M. (2019). On-line vis-NIR spectroscopy prediction of soil organic carbon using machine learning. *Soil and Tillage Research* 190, 120–127. https://doi.org/10.1016/J.STILL.2019.03.006.

Nocita, M., Stevens, A., Noon, C. and van Wesemael, B. (2013). Prediction of soil organic carbon for different levels of soil moisture using Vis-NIR spectroscopy. *Geoderma* 199, 37–42. https://doi.org/10.1016/j.geoderma.2012.07.020.

Nocita, M., Stevens, A., Toth, G., Panagos, P., van Wesemael, B. and Montanarella, L. (2014). Prediction of soil organic carbon content by diffuse reflectance spectroscopy using a local partial least square regression approach. *Soil Biology and Biochemistry* 68, 337–347. https://doi.org/10.1016/j.soilbio.2013.10.022.

Nouri, M., Gomez, C., Gorretta, N. and Roger, J. M. (2017). Clay content mapping from airborne hyperspectral Vis-NIR data by transferring a laboratory regression model. *Geoderma* 298, 54–66. https://doi.org/10.1016/J.GEODERMA.2017.03.011.

Oldfield, E. E., Bradford, M. A. and Wood, S. A. (2019). Global meta-analysis of the relationship between soil organic matter and crop yields. *SOIL* 5(1), 15–32. https://doi.org/10.5194/soil-5-15-2019.

Poggio, L., Gimona, A. and Brewer, M. J. (2013). Regional scale mapping of soil properties and their uncertainty with a large number of satellite-derived covariates. *Geoderma* 209–210, 1–14. https://doi.org/10.1016/J.GEODERMA.2013.05.029.

Ramirez-Lopez, L., Schmidt, K., Behrens, T., van Wesemael, B., Demattê, J. A. M. and Scholten, T. (2014). Sampling optimal calibration sets in soil infrared spectroscopy. *Geoderma* 226–227, 140–150. https://doi.org/10.1016/J.GEODERMA.2014.02.002.

Rodionov, A., Pätzold, S., Welp, G., Pallares, R. C., Damerow, L. and Amelung, W. (2014). Sensing of soil organic carbon using visible and near-infrared spectroscopy at variable moisture and surface roughness. *Soil Science Society of America Journal* 78(3), 949–957. https://doi.org/10.2136/sssaj2013.07.0264.

Rodionov, A., Pätzold, S., Welp, G., Pude, R. and Amelung, W. (2016). Proximal field Vis-NIR spectroscopy of soil organic carbon: A solution to clear obstacles related to vegetation and straw cover. *Soil and Tillage Research* 163, 89–98. https://doi.org/10.1016/J.STILL.2016.05.008.

Rogge, D., Bauer, A., Zeidler, J., Mueller, A., Esch, T. and Heiden, U. (2018). Building an exposed soil composite processor (SCMaP) for mapping spatial and temporal characteristics of soils with Landsat imagery (1984–2014). *Remote Sensing of Environment* 205, 1–17. https://doi.org/10.1016/j.rse.2017.11.004.

Sanderman, J., Hengl, T. and Fiske, G. J. (2017). Soil carbon debt of 12,000 years of human land use. *Proceedings of the National Academy of Sciences of the United States of America* 114(36), 9575–9580. https://doi.org/10.1073/pnas.1706103114.

Schaepman, M. E., Jehle, M., Hueni, A., D'Odorico, P., Damm, A., Weyermann, J., Schneider, F. D., Laurent, V., Popp, C., Seidel, F. C., Lenhard, K., Gege, P., Küchler, C., Brazile, J., Kohler, P., de Vos, L., Meuleman, K., Meynart, R., Schläpfer, D., Kneubühler, M. and Itten, K. I. (2015). Advanced radiometry measurements and earth science applications with the Airborne Prism Experiment (APEX). *Remote Sensing of Environment* 158, 207–219. https://doi.org/10.1016/J.RSE.2014.11.014.

Shenk, J. S. and Westerhaus, M. O. (1991). Population definition, sample selection and calibration procedures for Near infrared spectroscopy. *Crop Science* 31(2), 469–474.

Shepherd, K. D. and Walsh, M. G. (2007). Infrared spectroscopy—enabling an evidence-based diagnostic surveillance approach to agricultural and environmental management in developing countries. *Journal of Near Infrared Spectroscopy* 15(1), 1–19. https://doi.org/10.1255/jnirs.716.

Shi, P., Castaldi, F., van Wesemael, B. and van Oost, K. (2020). Large-scale, high-resolution mapping of soil aggregate stability in croplands using APEX hyperspectral imagery. *Remote Sensing* 12(4). https://doi.org/10.3390/rs12040666.

Stenberg, B., Viscarra Rossel, R. A., Mouazen, A. M. and Wetterlind, J. (2010). Visible and near infrared spectroscopy in soil science. In: *Advances in Agronomy* 107(C), 163–215. https://doi.org/10.1016/S0065-2113(10)07005-7.

Stevens, A., Udelhoven, T., Denis, A., Tychon, B., Lioy, R., Hoffmann, L. and van Wesemael, B. (2010). Measuring soil organic carbon in croplands at regional scale using airborne imaging spectroscopy. *Geoderma* 158(1-2), 32–45. https://doi.org/10.1016/j.geoderma.2009.11.032.

Tóth, G., Jones, A. and Montanarella, L. (2013). The LUCAS topsoil database and derived information on the regional variability of cropland topsoil properties in the European Union. *Environmental Monitoring and Assessment* 185(9), 7409–7425. https://doi.org/10.1007/s10661-013-3109-3.

Vaudour, E., Gomez, C., Loiseau, T., Baghdadi, N., Loubet, B., Arrouays, D., Ali, L. and Lagacherie, P. (2019). The impact of acquisition date on the prediction performance of topsoil organic carbon from Sentinel-2 for croplands. *Remote Sensing* 11(18). https://doi.org/10.3390/rs11182143.

Viscarra Rossel, R. A., Behrens, T., Ben-Dor, E., Brown, D. J., Demattê, J. A. M., Shepherd, K. D., Shi, Z., Stenberg, B., Stevens, A., Adamchuk, V., Aïchi, H., Barthès, B. G., Bartholomeus, H. M., Bayer, A. D., Bernoux, M., Böttcher, K., Brodský, L., Du, C. W., Chappell, A., Fouad, Y., Genot, V., Gomez, C., Grunwald, S., Gubler, A., Guerrero, C., Hedley, C. B., Knadel, M., Morrás, H. J. M., Nocita, M., Ramirez-Lopez, L., Roudier, P., Campos, E. M. R., Sanborn, P., Sellitto, V. M., Sudduth, K. A., Rawlins, B. G., Walter, C., Winowiecki, L. A., Hong, S. Y. and Ji, W. (2016). A global spectral library to characterize the world's soil. *Earth-Science Reviews* 155, 198–230. https://doi.org/10.1016/j.earscirev.2016.01.012.

Vos, C., Don, A., Hobley, E. U., Prietz, R., Heidkamp, A. and Freibauer, A. (2019). Factors controlling the variation in organic carbon stocks in agricultural soils of Germany. *European Journal of Soil Science* 70(3), 550–564. https://doi.org/10.1111/ejss.12787.

Ward, K. J., Chabrillat, S., Neumann, C. and Foerster, S. (2019). A remote sensing adapted approach for soil organic carbon prediction based on the spectrally clustered LUCAS soil database. *Geoderma* 353, 297–307. https://doi.org/10.1016/j.geoderma.2019.07.010.

Wijewardane, N. K., Ge, Y., Wills, S. and Loecke, T. (2016). Prediction of soil carbon in the conterminous United States: visible and near infrared reflectance spectroscopy analysis of the rapid carbon assessment project. *Soil Science Society of America Journal* 80(4), 973–982. https://doi.org/10.2136/sssaj2016.02.0052.

Xiao, J., Chevallier, F., Gomez, C., Guanter, L., Hicke, J. A., Huete, A. R., Ichii, K., Ni, W., Pang, Y., Rahman, A. F., Sun, G., Yuan, W., Zhang, L. and Zhang, X. (2019). Remote sensing of the terrestrial carbon cycle: a review of advances over 50 years. *Remote Sensing of Environment* 233, 111383. https://doi.org/10.1016/J.RSE.2019.111383, 111383.

Yue, J., Tian, J., Tian, Q., Xu, K. and Xu, N. (2019). Development of soil moisture indices from differences in water absorption between shortwave-infrared bands. *ISPRS Journal of Photogrammetry and Remote Sensing* 154, 216–230. https://doi.org/10.1016/J.ISPRSJPRS.2019.06.012.

Zacharias, S., Bogena, H., Samaniego, L., Mauder, M., Fuß, R., Pütz, T., Frenzel, M., Schwank, M., Baessler, C., Butterbach-Bahl, K., Bens, O., Borg, E., Brauer, A., Dietrich, P., Hajnsek, I., Helle, G., Kiese, R., Kunstmann, H., Klotz, S., Munch, J. C., Papen, H., Priesack, E.,

Schmid, H. P., Steinbrecher, R., Rosenbaum, U., Teutsch, G. and Vereecken, H. (2011). A network of terrestrial environmental observatories in Germany. *Vadose Zone Journal* 10(3), 955-973. https://doi.org/10.2136/vzj2010.0139.

Zhao, X., Xue, J. F., Zhang, X. Q., Kong, F. L., Chen, F., Lal, R. and Zhang, H. L. (2015). Stratification and storage of soil organic carbon and nitrogen as affected by tillage practices in the North China Plain. *PLoS ONE* 10(6), e0128873. https://doi.org/10.1371/journal.pone.0128873.

# Chapter 5

# Assessing the benefits of temperate agroforestry in enhancing carbon sequestration

*Augustine K. Osei[1] and Maren Oelbermann, University of Waterloo, Canada*

## 1 Introduction: moving from conservation to carbon - a historical overview

Cultivating agricultural crops in combination with trees dates back to the beginning of plant and animal domestication (King, 1987). Since then, numerous agroforestry practices have been developed throughout tropical and temperate environments. For example, the *Dehesa* or *Montado* system in Europe's Mediterranean region, which integrates cork (*Quercus suber* L.) and holm oak (*Quercus ilex* L.) with livestock grazing and cereal cultivation, has been used for ~4500 years (Nerlich et al., 2013). Although wood pastures, the combination of forests with pastures and field crops, date to Neolithic times in northern Europe, this farming system was still widely used in this region until the early twentieth century (King, 1987; Nerlich et al., 2013). Other common European agroforestry systems included orchard intercropping, hedgerows, and windbreaks (Nerlich et al., 2013). Prior to North American colonization, First Nations were active land managers by using fire to improve wildlife forage, hunt, and travel and boost medicinal plant and berry production (Castleden et al., 2009; Hoffman, 2022). In southwestern USA and northern Mexico, the soil

1 Corresponding author (a2osei@uwaterloo.ca).

beneath mesquite (*Prosopis* spp.), a nitrogen-fixing tree, was used to fertilize crops (Bainbridge et al., 1990). However, European settlement of North America introduced new agricultural methods, but some of the European agroforestry practices like tree-based intercropping, orchard intercropping, wood pastures, windbreaks, and home gardens were adopted in the new world (Williams et al., 1997). Furthermore, China has a long history of practicing agroforestry, where Han Dynasty (206 BC to AD 220) administrators recommended to develop forested areas that also included crop and livestock production (You, 1991). Thus, the frequency and distribution of woody vegetation throughout the temperate zone, on a global scale, was tipped toward tree species suitable and useful for human use, which caused trees to become an integral part of the agricultural landscape (Worms, 2021).

While human societies evolved and modernized so did food production systems. This also changed agricultural practices which became increasingly mechanized and intensified, causing trees to disappear from the landscape (Dupraz et al., 2018). In parallel, new policies and economic opportunities through increasing crop yields, combined with the development of agrochemicals, synthetic fertilizes, and crop breeding in the mid-twentieth century, transformed agroforestry practices to large-scale single-crop systems (Dupraz et al., 2018). Trees were considered a nuisance since they obstructed large machinery in the modern agricultural landscape. Furthermore, agroforestry wood pastures did not provide adequate nutritional requirements for modern livestock breeds and were converted to forests (Luick, 2009). This created a separation between agriculture and forestry (Reeg et al., 2008).

In the early twentieth century, intensified agricultural management practices contributed to soil compaction, erosion, loss of soil organic carbon (SOC), and soil fertility. Agricultural techniques and technologies brought by the Europeans were not suitable for the soils of the new world. This was the driving force behind the 1930s Dust Bowl in mid-western USA. Similar consequences of agricultural intensification and soil degradation occurred in other temperate regions in the Canadian prairies (Marchildon, 2009), Patagonia, Argentina (Peri et al., 2018), Chile (Dube et al., 2018), Australia (Reid and Moore, 2008), New Zealand (Kemp et al., 2018), China (Chang et al., 2018), Himalayan region of India (Kumar et al., 2018), and Europe (Dupraz et al., 2018). As a result, modern agriculture played a significant role in increasing the global total of marginal and degraded land that is now substandard for long-term food production (Amundson, 2022).

Due to increasing environmental degradation associated with single-crop systems, a resurgence of agroforestry practices occurred in the early 1970s. The renewed interest in agroforestry led to knowledge of which agroforestry practices are most effective for specific environmental conditions (King, 1987). However, King (1987) also noted that at this time research was scant and uncoordinated with a lack of reliable field data. Once the International Centre for Research on Agroforestry (World Agroforestry Centre) was established in 1977, organized and coordinated research began (King, 1987). Since then, our understanding of the soil–microbe–crop–atmospheric continuum advanced, and so did our understanding of the dynamic interactions when trees were integrated into the agricultural landscape (Nair et al., 2009). For example, newly developed agroforestry practices like alley cropping allowed for the systematic integration of trees into the cropped landscape (King, 1987). This provided an opportunity for agroforestry to move into the modern era of highly mechanized agriculture and trees began to be viewed as providing environmental services.

Although modernized agroforestry systems addressed multiple environmental issues related to soil and water conservation (Young, 1997), restoration of marginal/degraded lands (Montagnini et al., 2011), and improved yield (Pardon et al., 2018), intensified activities of tropical deforestation raised concerns on the impact of these activities on the global carbon cycle and climate change (Sánchez, 2000). To address this growing global issue, Dixon (1995) suggested that agroforestry systems could be used to rehabilitate marginal/degraded lands to sequester carbon and help mitigate climate change. Dixon (1995) estimated that for each hectare of agroforestry production, up to 20 ha of deforestation may be prevented. In fact, Dixon et al. (1993) suggested that the potential of agroforestry systems to store carbon was between 15 Mg C ha$^{-1}$ and 198 Mg C ha$^{-1}$, whereas Schroeder (1994) estimated 63 Mg C ha$^{-1}$, in temperate regions. The wide range of the potential of carbon sequestration in agroforestry systems suggested by Dixon et al. (1993) was due to regional differences, species composition, tree age, local environmental factors, and management practices of the agroforestry system. Consequently, agroforestry was integrated into the Kyoto Protocol due to its capacity to sequester carbon. Indeed, the 2000 IPCC report on land use, land-use change, and forestry recognized agroforestry systems as having the greatest potential for carbon sequestration of all land-use types (Watson et al., 2000). This is because trees function as an active carbon sink for many years by assimilating carbon in stems, branches, and roots until they die (Nair et al., 2009).

Agroforestry systems typically use fast-growing tree species, some of which have nitrogen-fixing capabilities (Nair et al., 2009). This enhances nutrient cycling and organic matter input from leaves/needles in addition to crop residues, causing a greater accumulation of SOC compared to single-crop agroecosystems that rely on carbon input from crop residues only (Oelbermann et al., 2004). For example, a 13-year-old hybrid poplar (*Populus* × *canadensis*) alley crop in southern Canada had an annual input of 537 g C m$^{-2}$ compared to a 480 g C m$^{-2}$ in a row crop system (Oelbermann et al., 2005). In the alley crop, 117 g C m$^{-2}$ was derived from autumnal litterfall from the 13-year-old trees. Furthermore, Oelbermann et al. (2005) suggested that this quantity of carbon input from autumnal litterfall increases with increasing tree age. Thus, agroforestry systems create a synergistic interaction between the soil, crops, trees, and the atmosphere to help sustain agricultural productivity with the co-benefit of continuously sequestering carbon for the purpose of climate mitigation (Oelbermann et al., 2004). However, until 30 years ago, the majority of information based on the carbon sequestration potential of agroforestry systems (c.f., Dixon et al., 1993, Schroeder, 1993) was based on models and the use of allometric equations (Kürsten and Burschel, 1993). But the capacity of agroforestry systems to sequester carbon, based on actual field data, from temperate regions was missing. Research output based on field data began to appear ~20 years ago (Oelbermann et al., 2004). Since then, numerous publications on carbon sequestration in various temperate agroforestry systems have appeared (Peichl et al., 2006; Bambrick et al., 2010; Udawatta and Jose, 2012; Wotherspoon et al., 2014; Mayer et al., 2022). However, Udawatta et al. (2022) found that our current understanding of carbon sequestration of various temperate agroforestry practices still remains rudimentary. Therefore, the objective of this chapter is to provide an overview of the current state of research and challenges associated with carbon sequestration studies in above and belowground tree components, in soil, and the long-term stabilization of carbon in soil in agroforestry systems commonly adopted in temperate environments. We will conclude by providing insights into future research directions.

## 2   System-level carbon dynamics in temperate agroforestry

At the system-level, agroforestry systems store carbon in above- and belowground components (Oelbermann et al., 2006). Aboveground components include standing biomass in woody stems and branches and biomass returned to the soil from crop residues and woody litter, whereas belowground components include woody and crop roots and soil humus (Fig. 1) (Oelbermann et al., 2006; Borden et al., 2014). Integrating woody and non-woody crops on the same land unit in agroforestry systems allows for more efficient capturing of growth resources for increased biomass growth and subsequent increase in biomass and soil carbon sequestration (Nair and Nair, 2003; Lorenz and Lal, 2018). Moreover, both woody and non-woody components in agroforestry systems contribute to carbon accumulation in above- and belowground systems (Nair et al., 2009). In this respect, agroforestry systems accumulate more carbon than sole-cropped systems, woody plantations, or pasture systems without woody components (Nair et al., 2009). Sharrow and Ismail (2004) observed 8.17 Mg ha$^{-1}$ year$^{-1}$ and 5.77 Mg ha$^{-1}$ year$^{-1}$ greater system-level carbon in an 11-year-old Douglas fir (*Pseudotsuga menziesii*) silvopasture system than in pure stands

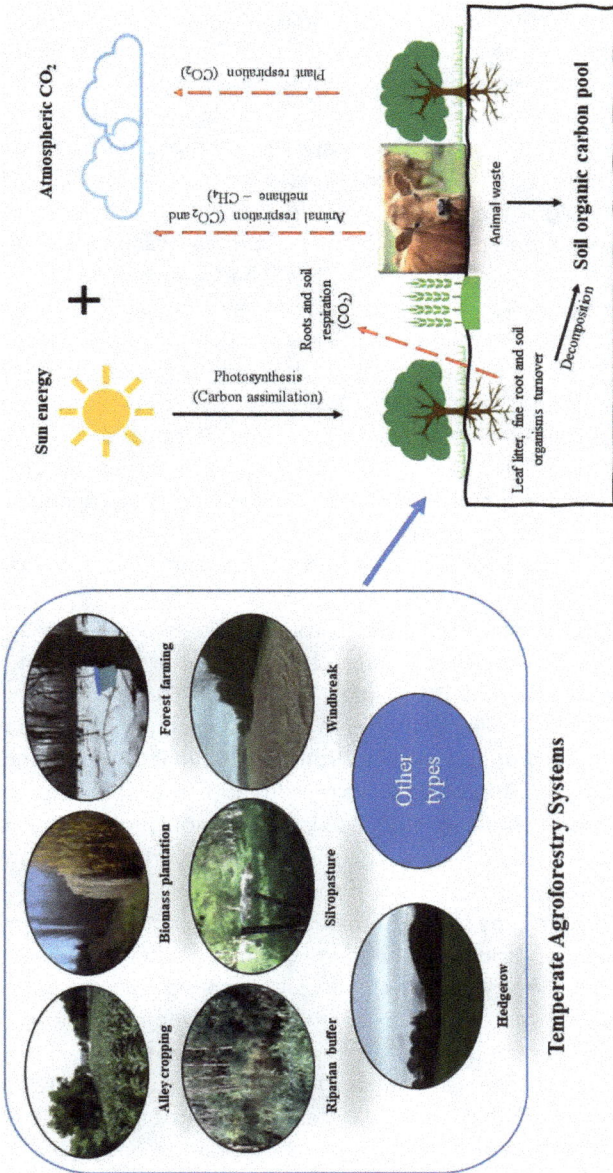

**Figure 1** A schematic representation of carbon cycle in agroforestry systems.

of Douglas-fir and pasture. Gordon and Thevathasan (2004) discovered that in southern Canada, silvopastoral systems incorporating fast-growing tree species, such as hybrid poplar, can sequester 2.7 to 3 times more carbon at the system-level than monoculture pasture systems. In a 13-year-old alley cropping system in southern Canada, Peichl et al. (2006) identified a system-level total carbon pool of 96.5 Mg C ha$^{-1}$ and 75.3 Mg C ha$^{-1}$ to a 20 cm soil depth, using hybrid poplar and Norway spruce, respectively, in the agroforestry system, in comparison to 68.5 Mg C ha$^{-1}$ in a barley sole crop. Similarly, Wotherspoon et al. (2014) determined system-level total carbon pools of 113.4, 99.4, 99.2, 91.5, and 91.3 Mg C ha$^{-1}$ in a 25-year-old alley cropping system in southern Canada, utilizing hybrid poplar, white cedar, red oak, black walnut, and Norway spruce, respectively, compared to 71.1 Mg C ha$^{-1}$ in the sole crop.

Unlike in tropical agroforestry systems, research on carbon dynamics in temperate agroforestry systems is relatively recent and still evolving (Chang et al., 2018). Carbon sequestration which refers to the capturing, removing, and storing of atmospheric carbon dioxide ($CO_2$) in long-term reservoirs such as terrestrial systems (e.g. vegetation and soil) (Fig. 1) is crucial to climate change mitigation (Nair et al., 2009). This natural process of capturing and storing carbon in terrestrial ecosystems is more cost-effective and sustainable compared to technological approaches (e.g. using engineering techniques to capture and inject industrial $CO_2$ into the ocean and deep geological strata) (Lal, 2008).

Studies evaluating system-level carbon stocks in temperate agroforestry systems are few. This may be because trees are generally thought to contain lower carbon than the soil (Peichl et al., 2006). Also, carbon assimilated in tree biomass is relatively labile compared to the soil carbon pool and may depend on the fate of the products derived from tree biomass (Upson and Burgess, 2013; Lorenz and Lal, 2018). For these reasons and other challenges associated with quantifying biomass carbon pools, the majority of carbon studies in temperate agroforestry systems largely focus on soil carbon with comparatively few studies quantifying biomass carbon.

In forest systems, the soil is estimated to contain about 60% of the system's carbon with 30% in aboveground parts (Nair et al., 2010). However, woody biomass in agroforestry systems can contribute significantly to the overall system's carbon sequestration potential. For example, a meta-analysis by Drexler et al. (2021) revealed that hedgerows in temperate regions stored 104 Mg ha$^{-1}$ more carbon than croplands. Out of this amount, biomass contributed 87 Mg ha$^{-1}$, representing 84% of the system's total carbon stock. Ma et al. (2020) also found that global agroforestry systems, on average, had 46.1 Mg ha$^{-1}$ more carbon in tree biomass than crop and pastureland without trees. This suggests that at the system-level, biomass carbon accumulation plays a vital contribution to the carbon sequestration potential of agroforestry systems.

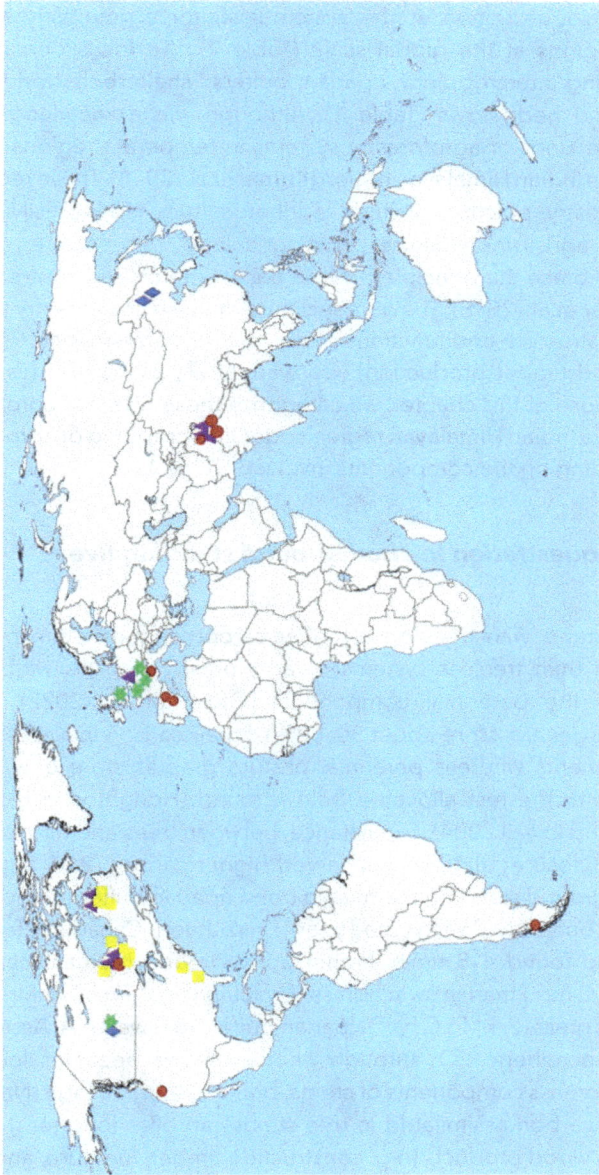

**Figure 2** The global distribution of the different temperate agroforestry systems in the study sites used for our analysis of carbon sequestration rates in this study.

Legend: ▲ Alley cropping/tree intercropping; ✳ Hedgerows; ◼ Riparian buffers; ◆ Shelterbelts/windbreaks; and ● Silvopastures

To understand the carbon sequestration benefits of temperate agroforestry systems, we review studies (Fig. 2; Table 1) that quantify carbon stocks in aboveground standing biomass, belowground biomass (roots), and soil and analyze carbon sequestration at the system-level for agroforestry systems in temperate regions at the global scale (Table 2). We focus our analyses on alley cropping/intercropping, riparian buffers, shelterbelts/windbreaks, silvopastures, and hedgerows (Table 1). Although, we acknowledge there are other diverse kinds of agroforestry systems in temperate regions in Asia, particularly, in the Indian Himalayan region (Kumar et al., 2018). These regionally specific agroforestry practices include agrihorticulture, agrihortisilviculture, agrisilviculture, agrisilvihorticulture, hortiagriculture, silvipasture, among others, which embrace the principles of tree-based intercrops with crops and/or pasture (Kumar et al., 2018). The nomenclature of these agroforestry systems is based on the structure and functions of the various components (e.g. food, fruits, fodder, and timber production) (Goswani et al., 2014). For this reason and for the purpose of this chapter, we categorize these different agroforestry practices from the Indian Himalayan region under intercropping or silvopasture systems, depending on the components involved (Table 1).

## 2.1 Carbon sequestration in aboveground standing live biomass

Agroforestry systems with trees have higher carbon sequestration potential in aboveground than treeless systems or agroforestry systems with shrubs and grasses as the perennial components (Ofosu et al., 2022). This is because trees in general store about 50–60% of their carbon in aboveground biomass components whereas perennial pasture grasses store only 10% in aboveground, with the rest allocated belowground (Houghton and Hackler 2000; Sharrow and Ismail 2004). For instance, between trees and riparian grass buffer systems, Ofosu et al. (2022) observed higher carbon (2.31 Mg C ha$^{-1}$ year$^{-1}$) in aboveground standing biomass in treed riparian buffers compared to grassed riparian buffers (0.05 Mg C ha$^{-1}$ year$^{-1}$) in southern Canada. Tufekcioglu et al. (2003) also found 4–8 times higher aboveground biomass carbon in poplar (~20 Mg C ha$^{-1}$) than in switchgrass (*Panicum virgatum* L.) (5 Mg C ha$^{-1}$) and cool-season grass (2.5 Mg C ha$^{-1}$) riparian buffers in Iowa, USA. Assimilated carbon from atmospheric $CO_2$ through photosynthesis becomes locked in aboveground biomass components of stems, branches, and foliage (Nair et al., 2009). Whereas carbon assimilated in tree stems can offer long-term carbon sequestration as wood products (e.g. construction timber, furniture, and wood crafts) (Nair et al., 2009; Udawatta et al., 2022), carbon from tree foliage can be stored over the long-term in the soil through litter input as soil organic matter (SOM) (Fig. 1) (Oelbermann et al., 2004).

**Table 1** System-level carbon stocks in different temperate agroforestry systems and adjacent land uses from different locations (only studies published in the past 20 years that included aboveground, belowground (root) and soil carbon pools were considered)

| Agroforestry system | Location | Age (Years) | Dominant tree/shrub species | Carbon stocks in agroforestry system (Mg ha⁻¹)* | | | Annual detrital carbon inputs (Mg ha⁻¹ year⁻¹) | Adjacent land use | Carbon stocks in adjacent land use (Mg ha⁻¹) | | Source |
| --- | --- | --- | --- | --- | --- | --- | --- | --- | --- | --- | --- |
| | | | | Aboveground | Belowground | Soil | | | Biomass | Soil | |
| **Agri-horticulture (Intercropping)** | Himachal Pradesh, India | NA† | Apple (*Malus domestica*), Japanese Persimmon (*Diospyros kaki*), pear (*Pyrus communis*) | 33.38 | 7.46 | 56.34 (0–30)‡ | 0.39 | NA | NA | NA | Bhardwaj et al. (2023) |
| | Northwestern Himalaya, India | NA | Mango (*Mangifera indica*), citrus (*Citrus* spp.), plum (*Prunus domestica*), peach (*Prunus persica*), pear | 13.94 | 3.52 | 22.73 (0–30) | NA | NA | NA | NA | Sharma et al. (2023) |
| | Northwestern Himalaya, India | 21 | Apple | 11.08 | 3.67 | 151.77 (0–100) | | Agriculture | 3.34 | 132.75 (0–100) | Chisanga et al. (2018) |
| | Himachal Pradesh, India | 24 | Not provided | 32.37 | 16.68 | 42.51 (0–40) | 1.26 | Agriculture | 12.22 | 34.72 (0–40) | Rajput et al. (2017) |
| | Himachal Pradesh, India | ~30 | Gooseberry (*Emblica officinalis*), citrus, peach | 9.36 | 2.79 | 90.07 (0–40) | | Agriculture | 8.24 | 65.43 (0–40) | Goswami et al. (2014) |
| **Agri-horti-silviculture (Intercropping)** | Himachal Pradesh, India | NA | Armenian plum (*Prunus armeniaca*), nettle tree (*Celtis australis*), oak (*Quercus ilex*), white willow (*Salix alba*), Himalayan poplar (*Populus ciliate*), Bhutan pine (*Pinus wallichiana*), Chilghoza pine (*Pinus gerardiana*) | 45.08 | 11.13 | 58.75 (0–30) | 0.42 | NA | NA | NA | Bhardwaj et al. (2023) |
| **Agri-silviculture (Intercropping)** | Himachal Pradesh, India | NA | Armenian plum, oak, Himalayan poplar, Chilghoza pine, apple | 48.66 | 14.44 | 52.06 (0–30) | 0.45 | NA | NA | NA | Bhardwaj et al. (2023) |
| | Northwestern Himalaya, India | NA | Australian red cedar/Toon tree (*Toona ciliata*), bhimal tree (*Grewia optiva*), orchid tree (*Bauhinia variegate*), red silk cotton tree (*Bombax ceiba*), *Albizia chinensis*, *Ficus palmata*), mulberry (*Morus alba*), Nettle tree, Himalayan poplar | 19.56 | 4.95 | 25.37 (0–30) | NA | NA | NA | NA | Sharma et al. (2023) |
| | Himachal Pradesh, India | ~30 | Bhimal tree, fig tree (*Ficus* spp.), Nettle tree | 8.19 | 2.45 | 92.65 (0–40) | | Agriculture | 8.24 | 65.43 (0–40) | Goswami et al. (2014) |
| **Agri-silvihorti-culture (Intercropping)** | Himachal Pradesh, India | ~30 | Nettle tree, bhimal, fig tree, Australian red cedar/toon tree, pomegranate (*Punica granatum*), Armenian plum | 11.38 | 3.40 | 95.46 (0–40) | | Agriculture | 8.24 | 65.43 (0–40) | Goswami et al. (2014) |

*(Continued)*

**Table 1** (*Continued*)

| Agroforestry system | Location | Age Years | Dominant tree/shrub species | Carbon stocks in agroforestry system (Mg ha⁻¹)* | | | Annual detrital carbon inputs Mg ha⁻¹ year⁻¹ | Adjacent land use | Carbon stocks in adjacent land use (Mg ha⁻¹) | | Source |
|---|---|---|---|---|---|---|---|---|---|---|---|
| | | | | Aboveground | Belowground | Soil | | | Biomass | Soil | |
| **Agro-horti-silviculture** (*Intercropping*) | Northwestern Himalaya, India | NA | Himalayan poplar, nettle tree, deodar cedar (*Cedrus deodara*), Indian horse chestnut (*Aesculus indica*), Indian willow (*Salix tetrasperma*), apple, plum, pear, English walnut (*Juglans regia*) | 23.74 | 6.10 | 25.92 (0–30) | NA | NA | NA | NA | Sharma et al. (2023) |
| **Horti-agriculture** (*Intercropping*) | Northwestern Himalaya, India | NA | Apple, pear, hazel (*Corylus jacquemontii*), English walnut | 18.04 | 4.69 | 22.25 (0–30) | NA | NA | NA | NA | Sharma et al. (2023) |
| **Agro-silvohorti-culture** (*Intercropping*) | Northwestern Himalaya, India | | Australian red cedar tree, bhimal tree, Albizia tree (*Albizia chinensis*), Chinaberry tree (*Melia composita*) *Leucaena leucocephala*, mango, citrus, pear, plum, peach | 17.87 | 4.48 | 31.66 (0–30) | NA | NA | NA | NA | Sharma et al. (2023) |
| **Alley cropping/ Intercropping Systems** | Northern Himalaya, India | 11 | Apple | 14.31 | 3.82 | 45.15 (0–20) | NA | Agriculture | 4.58 | 34.94 (0–20) | Zahoor et al. (2021) |
| | France | 6 | Hybrid walnut (*Juglans × intermedia* Carr.) | 0.02 | 0.01 | 96.10 (0–30) | NA | Agriculture | NA | 90.90 (0–30) | Cardinael et al. (2017) |
| | | 6 | Hybrid walnut | 0.07 | 0.03 | 82.70 (0–30) | NA | Agriculture | NA | 78.70 (0–30) | |
| | | 41 | Black walnut (*Juglans nigra*) | 19.85 | 5.55 | 96.40 (0–20) | NA | Agriculture | NA | 65.10 (0–20) | |
| | | 18 | Hybrid walnut | 26.64 | 6.61 | 289.40 (0–60) | NA | Agriculture | NA | 263.80 (0–60) | |
| | | 18 | Hybrid walnut | 10.88 | 2.99 | 317.90 (0–100) | NA | Agriculture | NA | 294.30 (0–100) | |
| | Southern Quebec, Canada | 10 | Hybrid poplar (*Populus × canadensis*), red oak (*Quercus rubra* L.), and black cherry (*Prunus serotina* Ehrh.) | 31.09 | 7.25 | 148.50 (0–30) | 0.04 | Agriculture | 2.28 | 131.90 (0–30) | Winans et al. (2016) |
| | | 10 | Hybrid poplar, red oak, and white ash (*Fraxinus americana* L.) | 70.75 | 15.95 | 159.90 (0–30) | 0.06 | Non-tree-based forage | 3.19 | 167.90 (0–30) | |

| Location | Age | Species | | | | | Land use | | | Reference |
|---|---|---|---|---|---|---|---|---|---|---|
| Himachal Pradesh, India | 20 | Apple and plum | 36.42 | 11.16 | 41.87 (0-40) | 2.38 | Agriculture | 15.08 | 35.77 (0-40) | Rajput et al. (2015) |
| Southern Ontario, Canada | 25 | Poplar (*Populus* spp.) | 21.06 | 5.94 | 86.86 (0-40) | 2.45 | Agriculture | 1.40 | 71.08 (0-40) | Wotherspoon et al. (2014) |
| | 25 | Red oak | 11.84 | 4.16 | 83.77 (0-40) | 1.60 | | | | |
| | 25 | Black walnut | 12.30 | 2.70 | 76.84 (0-40) | 2.25 | | | | |
| | 25 | Norway spruce (*Picea abies* L.) | 9.49 | 3.51 | 78.33 (0-40) | 1.94 | | | | |
| | 25 | White cedar (*Thuja occidentalis*) | 14.88 | 1.12 | 83.23 (0-40) | 0.88 | | | | |
| Southern Ontario, Canada | 13 | Hybrid poplar | 14.80 | 3.20 | 78.50 (0-20) | 1.20 | Agriculture | 3.50 | 65.00 (0-20) | Peichl et al. (2006) |
| | 13 | Norway spruce | 7.10 | 2.20 | 66.00 (0-20) | 0.12 | | | | |
| Northwestern Himalaya, India | 50 | Black locust (*Robinia* spp.), Apple | 30.25 | 8.14 | 125.37 (0-100) | | Agriculture | 3.34 | 132.75 (0-100) | Chisanga et al. (2018) |
| Himachal Pradesh, India | ~30 | Nettle tree, Pomegranate, Pear, Bhimal tree | 11.13 | 3.32 | 84.14 (0-40) | | Agriculture | 8.24 | 65.43 (0-40) | Goswami et al. (2014) |
| Western France | 20–120 | Chestnut (*Castanea sativa*), Elderberry, English oak (*Quercus robur*), Hornbeam (*Carpinus betulus*), Hazel (*Corylus avellana*), European ash (*Fraxinus excelcior*), Beech (*Fagus sylvatica*), Black locust, Rowan (*Sorbus aucuparia*), Elm (*Ulmus minor*), Maple (*Acer campestre*), Hawthorn (*Crataegus monogyna*) | 1.20–21.60 Mg C 100 m⁻¹ | 0.30–6.10 Mg C 100 m⁻¹ | 109.05–155.4 (0-90) | NA | NA | NA | NA | Viaud and Kunnemann (2021) |
| **Hedgerows** Central Alberta, Canada | 27 | Wood rose (*Rosa woodsii* Lindl.), Raspberry (*Rubus idaeus* L.), Aspen (*Populus tremuloides* Michx.), Balsam poplar (*Populus balsamifera* L.) | 126.49 | 35.51 | 315.79 (0-100) | 1.64 | Agriculture | 3.85 | 168.30 (0-100) | Gross et al. (2022) |
| Southeast England | 100 | Blackthorn (*Prunus spinosa*) | 600.00 | 110.00 | 210.00 (0-50) | 35.04 | NA | NA | NA | Crossland (2015) |
| | 100 | Hawthorn | 179.00 | 50.00 | 990.00 (0-50) | 32.30 | NA | NA | NA | |
| | 100 | Hazel | 125.00 | 49.00 | 100.00 (0-50) | 20.85 | NA | NA | NA | |

*(Continued)*

**Table 1** (Continued)

| Agroforestry system | Location | Age (Years) | Dominant tree/shrub species | Carbon stocks in agroforestry system (Mg ha⁻¹)* | | | Annual detrital carbon inputs (Mg ha⁻¹ year⁻¹) | Adjacent land use | Carbon stocks in adjacent land use (Mg ha⁻¹) | | Source |
|---|---|---|---|---|---|---|---|---|---|---|---|
| | | | | Aboveground | Belowground | Soil | | | Biomass | Soil | |
| **Riparian buffers** | Southern Ontario, Canada | 50 | Norway spruce | 184.60 | 55.30 | 271.80 (0–60) | 2.30 | Agriculture | NA | 113.60 (0–60) | Ofosu et al. (2022) |
| | | 156 | White cedar, sugar maple (*Acer saccharinum*), birch (*Betula alleghaniensis*), black walnut | 207.70 | 62.10 | 658.30 (0–60) | 2.70 | Agriculture | NA | 101.30 (0–60) | |
| | | 156 | Hybrid poplar White cedar, BirchAsh | 276.30 | 82.70 | 578.30 (0–60) | 3.60 | Agriculture | NA | 157.10 (0–60) | |
| | | 32 | Sugar maple, black walnut, hybrid poplar | 243.10 | 72.70 | 356.90 (0–60) | 4.50 | Agriculture | NA | 157.10 (0–60) | |
| | | 35 | Bentgrass (*Agrostis* spp.) and purple-stemmed aster (*Symphyotrichum puniceum*). | 1.70 | 3.20 | 286.00 (0–60) | 0.90 | Agriculture | NA | 164.90 (0–60) | |
| | Southern Ontario, Canada | 37–60 | Sugar maple and Ash | 171.78 | 73.62 | 229.61 (0–30) | NA | Agriculture | NA | 97.40 (0–30) | Vijayakumar et al. (2020) |
| | | 34–60 | White cedar and white pine (*Pinus strobus*) | 137.76 | 59.04 | 155.75 (0–30) | NA | Agriculture | NA | 78.95 (0–30) | |
| | | 6–8 | Sugar maple and Black walnut | 12.18 | 5.22 | 124.07 (0–30) | NA | Agriculture | NA | 118.45 (0–30) | |
| | | 6–8 | White spruce and white pine | 4.90 | 2.10 | 131.41 (0–30) | NA | Agriculture | NA | 100.00 (0–30) | |
| | Southern Quebec, Canada | 9–200 | Eastern hemlock (*Tsuga canadensis*), white cedar, grey birch (*Betula populifolia*), sugar maple, hybrid poplar | 25.00–115.00 | 4.80–36.80 | 83.00–176.20 (0–60) | 2.30–12.30 | Pasture | 3.7 | 122.39 (0–60) | Fortier et al. (2013; 2015) |
| | North Carolina, USA | >50 | Black gum (*Nyssa biflora*), red oak, sweetgum (*Liquidambar styraciflua*), red maple (*Acer rubrum*) | 115.66 | 40.64 | 39.15 (0–10) | 0.93 | Agriculture | 0.85 | 16.25 (0–10) | Rheinhardt et al. (2012) |
| | | 25–50 | Red maple, sweetgum, loblolly pine (*Pinus taeda*) | 44.51 | 15.64 | 33.40 (0–10) | 0.70–1.39 | | | | |
| | | 5–25 | loblolly pine, sweetgum, red maple, black willow (*Salix nigra*) | 45.40 | 15.95 | 28.55 (0–10) | 0.50–2.48 | | | | |
| | | 0–5 | | 1.63 | 0.57 | 41.45 (0–10) | 7.76 | | | | |

| Location | Age | Species | | | | | Previous land use | | | Reference |
|---|---|---|---|---|---|---|---|---|---|---|
| Northeastern Ontario, Canada | 95 | Black spruce (Picea mariana), white spruce (Picea glauca), balsam fir (Abies balsamea), whitebirch (Betula Papyrifera), mountain maple (Acer spicatum), speckled alder (Alnus rugosa), and beakedhazel (Corylus cornuta) | 63.56 | 20.29 | 52.65 (0–75) | 0.38 | Forest | 81.41 | 68.02 (0–75) | Hazlett et al. (2005) |
| South Carolina, USA | 5 | Blackberry (Rubus spp.), smooth alder (Alnus serrulata), and buttonbush (Cephalanths occidentalis L.) | 5.35 | 1.85 | 20.26 (0–10) | 0.07 | NA | NA | NA | Giese et al.(2003) |
| | 11 | Willow, Waxmyrtle (Myrica cerifera L.), smooth alder, and buttonbush | 17.14 | 2.77 | 17.47 (0–10) | 1.64 | NA | NA | NA | |
| | 15 | Willow, red maple, smooth alder, Waxmyrtle, river birch (Betula nigra L.), sweetgumand persimmon (Diospyros virginiana L.) | 17.23 | 3.04 | 15.84 (0–10) | 1.19 | NA | NA | NA | |
| | 63 | Matured bottomlandhardwood riparian forest with several mixed species | 101.77 | 4.36 | 12.04 (0–10) | 1.35 | NA | NA | NA | |
| **Shelterbelts** Central Alberta, Canada | 29 | Balsam poplar, and willow | 130.03 | 23.57 | 210.11 (0–100) | 0.46 | Agriculture | 4.53 | 175.96 (0–100) | Gross et al. (2022) |
| The Three-North, China | 25 | Mixed woodlands and shrublands | 11.19 | 12.93 | 70.32 | 0.17 | NA | NA | NA | Chu et al. (2019) |
| **Pastoral-silviculture (Silvopasture)** Northwestern Himalaya, India | NA | Himalayan poplar, nettle tree, deodar cedar (Cedrus deodara), pear | 13.89 | 3.47 | 28.15 (0–30) | NA | NA | NA | NA | Sharma et al. (2023) |
| **Horti-pastoral (Silvopasture)** Northwestern Himalaya, India | NA | Apple, pear, English walnut | 15.92 | 4.02 | 23.96 (0–30) | NA | NA | NA | NA | Sharma et al. (2023) |
| **Silvopasture systems** Northwestern Himalaya, India | NA | Australian red cedar, Albizia chinensis, orchid tree, nettle tree, pear, Ficus palmata, Mulberry, Chir pine (Pinusroxburghii), Zizyphus nummularia, Dodonea viscosa | 19.09 | 4.90 | 35.77 (0–30) | NA | NA | NA | NA | Sharma et al. (2023) |
| Northwestern Himalaya, India | 70 | Chilgoza pine (Pinus gerardiana) | 42.33 | 9.75 | 108.97 (0–100) | NA | Agriculture | 3.34 | 132.75 (0–100) | Chisanga et al. (2018) |
| France | 26 | Wild cherry (Prunus avium) | 36.69 | 9.13 | 541.70 (0–50) | NA | Agriculture | NA | 540.00 (0–50) | Cardinael et al. (2017) |
| Northern Extremadura, Spain | 13 | Hybrid walnut | 14.72 | 10.57 | 91.20 (0–100) | NA | Hybrid walnut plantation with ploughing | 26.13 | 81.49 (0–100) | López-Díaz et al. (2017) |

*(Continued)*

**Table 1** (*Continued*)

| Agroforestry system Location | Age Years | Dominant tree/shrub species | Carbon stocks in agroforestry system (Mg ha⁻¹)* | | | Annual detrital carbon inputs Mg ha⁻¹ year⁻¹ | Adjacent land use | Carbon stocks in adjacent land use (Mg ha⁻¹) | | Source |
|---|---|---|---|---|---|---|---|---|---|---|
| | | | Aboveground | Belowground | Soil | | | Biomass | Soil | |
| Himachal Pradesh, India | 52 | NA (silvipasture) | 40.70 | 20.96 | 38.54 (0–40) | 0.80 | Agriculture | 12.22 | 34.72 (0–40) | Rajput et al. (2017) |
| Himachal Pradesh, India | ~30 | Acacia (*Acacia catechu*), nettle tree, Chir pine, pear, Himmalayan oak | 2.59 | 0.77 | 115.45 (0–40) | | Agriculture | 8.24 | 65.43 (0–40) | Goswami et al. (2014) |
| | | | | | | | Grassland | 1.23 | 80.15 (0–40) | |
| Patagonia, Chile | 18 | Ponderosa pine (*Pinus ponderosa*) | 22.73 | 11.81 | 193.76 (0–40) | NA | *Pinus ponderosa* plantation | 47.74 | 149.25 (0–40) | Dube et al. (2012) |
| | | | | | | | Natural pasture | 4.65 | 177.10 (0–40) | |
| Spain | 11 | Monterey pine (*Pinus radiata*) | 73.60 | 19.33 | 138.25 (0–25) | 2.58 | NA | NA | NA | Fernández-Núñez et al. (2010) |
| | 11 | Silver birch | 29.99 | 4.17 | 157.21 (0–25) | NA | NA | NA | NA | |
| Southern Ontario, Canada | 15 | Hybrid poplar | 18.21 | 4.94 | 15.21 (0–5) | 1.92 | Ryegrass monoculture | 7.96 | 14.24 (0–5) | Gordon and Thevathasan (2004) |
| Oregon, USA | 11 | Douglas-fir (*Pseudotsuga menziesii*) | 10.71 | 1.53 | 95.89 (0–45) | 0.11 | Pasture | 1.00 | 102.52 (0–45) | Sharrow and Ismail (2004) |

†NA, not determined in study, hence data not available.
*Where C concentration was not provided, biomass C was assumed as 50% of total biomass.
‡Soil depth (cm) is provided in parentheses.

**Table 2** Estimated carbon sequestration rate in aboveground, belowground, soil, and at the system-level for different agroforestry systems in temperate regions

| Agroforestry system | Tree/shrub density Average No. of Tree/shrub ha⁻¹ | Agroforestry component | Carbon stocks Minimum ----------Mg C ha⁻¹---------- | Maximum | Average | Average age (Years†) | Carbon sequestration rate ----------Mg C ha⁻¹ year⁻¹---------- | Total system-level carbon ----------Mg C ha⁻¹ year⁻¹---------- |
|---|---|---|---|---|---|---|---|---|
| **Alley cropping/ Intercropping systems** | 240 | Aboveground | 0.02 | 70.75 | 20.50 | 24 | 0.85 | |
| | | Belowground (roots) | 0.01 | 16.68 | 5.60 | | 0.23 | |
| | | Soil | 22.25 | 317.90 | 90.60 | | 0.64 (0–50) 0.87 (0–100) | **1.72 (0–50)** **1.95 (0–100)** |
| **Riparian buffers** | 1424 | Aboveground | 1.63 | 276.30 | 89.61 | 49 | 1.83 | |
| | | Belowground (roots) | 0.57 | 82.70 | 28.13 | | 0.57 | |
| | | Soil | 12.04 | 658.30 | 165.61 | | 3.08 (0–50) 2.75 (0–100) | **5.48 (0–50)** **5.15 (0–100)** |
| **Shelterbelts** | 1200 | Aboveground | 11.19 | 130.03 | 70.61 | 27 | 2.62 | |
| | | Belowground (roots) | 12.93 | 23.57 | 18.25 | | 0.68 | |
| | | Soil | 70.32 | 210.11 | 140.22 | | 1.18 (0–100) | **4.48 (0–100)** |
| **Silvopasture systems** | 200 | Aboveground | 2.59 | 73.60 | 26.24 | 27 | 0.97 | |
| | | Belowground (roots) | 0.77 | 20.96 | 8.10 | | 0.30 | |
| | | Soil | 15.21 | 541.70 | 121.85 | | 0.45 (0–50) 0.20 (0–100) | **1.72 (0–50)** **1.47 (0–100)** |
| **Hedgerows** | 7533 | Aboveground | 125.00 | 600.00 | 257.62 | 82 | 3.14 | |
| | | Belowground (roots) | 35.51 | 110.00 | 61.13 | | 0.75 | |
| | | Soil | 100.00 | 990.00 | 403.95 | | 5.46 (0–100) | **9.35 (0–100)** |

†Average age for each system was calculated by averaging the years for that system as reported in Table 1. Where years are provided in range from Table 1, average years were calculated from the primary source of the study. Age for systems with NA were assumed as 30 years based on average age of similar agroforestry practices in the Indian Himalayan Region.

‡Annual SOC sequestration rates for each agroforestry system were calculated as:

$$\text{Annual SOC Sequestration Rate} = \left( \frac{SOC\ stocks_{Agroforestry} - SOC\ stocks_{Adjacent\ land\ use}}{Age\ of\ Agroforestry\ system} \right).\ \text{Soil depth (cm) is provided in parenthesis.}$$

*NA, Annual SOC sequestration rates could not be determined due to the absence of adjacent land use in the study

Temperate agroforestry systems are estimated to sequester 1.9 Pg C year$^{-1}$ in aboveground components (Oelbermann et al., 2004) but to date, only a few studies have quantified biomass carbon sequestration potentials of temperate agroforestry systems. The paucity of data on aboveground biomass carbon in temperate agroforestry systems may be attributed to challenges associated with tree biomass estimations (Sinacore et al., 2017). Estimating aboveground biomass is challenging because the most accurate approach for generating reliable biomass estimates involves destructive sampling and determining carbon concentrations in the various aboveground biomass components (Ketterings et al., 2001; Kaonga and Bayliss-Smith, 2009). This approach is, however, complex, time-consuming, destructive, and labor intensive which restricts its use to small areas and small tree sample sizes (Ketterings et al., 2001; Kaonga and Bayliss-Smith, 2009).

Due to the large number of tree species that are sometimes available in a particular agroforestry system and the challenges associated with destructively sampling large-size trees to accurately determine aboveground biomass, biomass regression models and sometimes remote sensing techniques such as LiDAR are used in estimating aboveground biomass (Drake et al., 2003; Sinacore et al., 2017). Also, in most of these studies, carbon is estimated based on the assumption that biomass contains 45–50% carbon (Goswani et al., 2014; Udawatta et al., 2022). Moreover, in some of these studies, biomass carbon of associated crops is often neglected. Given trees in agroforestry systems influence understory or companion crops in many diverse ways (Lorenz and Lal, 2018), neglecting carbon stocks in these secondary biomass components may result in a significant underestimation of the carbon sequestration potential of agroforestry systems.

Most of the studies that have quantified aboveground biomass carbon stocks in temperate agroforestry systems are concentrated in temperate North America and in the Indian Himalayan region (Table 1). Results from these studies are widely variable where the carbon sequestration potential of aboveground standing biomass ranges from 0.85 Mg C ha$^{-1}$ year$^{-1}$ in alley cropping/intercropping systems to 3.14 Mg C ha$^{-1}$ year$^{-1}$ in hedgerows (Table 2). Cardinael et al. (2018) determined that aboveground carbon sequestration rates, based on a global literature review for temperate agroforestry systems, ranges from 0.52 to 3.11 Mg C ha$^{-1}$ year$^{-1}$. Nair and Nair (2003) also determined that the aboveground biomass carbon sequestration rates for different agroforestry systems ranged from 0.29 to 15.21 Mg C ha$^{-1}$ year$^{-1}$. The wide range determined by Nair and Nair (2003) was attributed to differences in management, species, age, and differences in ecological and edaphic factors.

Udawatta et al. (2022) reported aboveground biomass carbon sequestration rates in alley crops (2.6 Mg ha$^{-1}$ year$^{-1}$), riparian buffers (3.0 Mg ha$^{-1}$ year$^{-1}$), and silvopasture (4.9 Mg ha$^{-1}$ year$^{-1}$) in temperate North America.

These aboveground biomass carbon rates are higher than the 0.9 Mg C ha$^{-1}$ year$^{-1}$, 1.8 Mg C ha$^{-1}$ year$^{-1}$, and 1.0 Mg C ha$^{-1}$ year$^{-1}$ that we calculate for alley crops, riparian buffers, and silvopastures, respectively, for temperate regions at the global scale (Table 2). Whereas the analysis by Udawatta et al. (2022) considered relatively matured agroforestry systems in only USA and Canada, our synthesis covers all global temperate regions including the dry and cold temperate region of the Indian Himalayas (e.g. cold deserts) that incorporate diverse tree species of various age classes and types of agroforestry systems.

Furthermore, local environmental and edaphic conditions, tree species, and age of the agroforestry system impact biomass productivity, and therefore carbon sequestration rates (Ma et al., 2020). Schroeder (1994) estimated that aboveground biomass carbon sequestration rate for temperate riparian buffers with 30-year rotation cycle was 2.1 Mg C ha$^{-1}$ year$^{-1}$. However, Udawatta and Jose (2012) estimated a carbon sequestration rate of 2.5 Mg C ha$^{-1}$ year$^{-1}$ for mature riparian buffers in temperate North America with 50-year rotation cycle. Vijayakumar et al. (2020) also found that mature (34–60 years) riparian buffers had a higher (1.1–4.1 Mg ha$^{-1}$ year$^{-1}$) carbon sequestration rate in aboveground biomass than younger (6–8 years) riparian buffer systems (0.3–1.1 Mg ha$^{-1}$ year$^{-1}$) in southern Canada.

In general, as an agroforestry system matures, the biomass of the woody components and understory vegetation increases, resulting in an overall increase in biomass carbon stocks (Udawatta and Jose, 2012). Since tree growth follows a sigmoidal pattern (Weiner and Thomas, 2001), the accumulation of biomass carbon is expected to occur with age until the tree reaches its maximum growth. However, the time taken for a tree to reach its full growth, and its biomass carbon sequestration rate, depends on the species. Balian and Naiman (2005) observed that ~90% of aboveground biomass accumulation occurred during the first 20–40 years in riparian buffer with willow (*Salix* spp.) and red alder (*Alnus rubra*) as the dominant species. Yet, when sitka spruce (*Picea sitchensis*) dominated the riparian buffer landscape, biomass carbon accrual by the system did not reach equilibrium until after 300 years. In a tree-based intercropping study in southern Canada, Wotherspoon et al. (2014) observed that fast-growing hybrid poplar reached its full carbon sequestration potential 25 years after establishment compared to the Norway spruce (*Picea abies*) which could continue to sequester atmospheric $CO_2$ until after 60 years. Nair et al. (2009) suggested the use of slow-growing rather than fast-growing tree species in agroforestry systems if carbon sequestration is the intended purpose of the agroforestry system. Apart from accumulating carbon over the long-term, slow-growing species have high specific-gravity wood and are deemed valuable for use as wood products (e.g. construction timber, furniture, and wood crafts) which constitute longer-term sink for fixed carbon (Nair et al., 2009).

Agroforestry systems such as riparian buffers and hedgerows tend to sequester more carbon in their biomass than systems which integrate annual row crops on the same land space such as alley cropping or intercropping (Thevathasan et al., 2018). We determine that hedgerows and riparian buffers have a 269% and 115% higher aboveground biomass carbon sequestration rate than alley cropping/intercropping system, respectively (Table 2). Thevathasan et al. (2018) suggested that riparian buffers have greater potential for aboveground biomass carbon sequestration than alley cropping systems because the absence of associated annual crops in riparian systems allows for a greater tree density. Also, since riparian plantings are located in the terrestrial-aquatic ecotone, competition for moisture is normally not a limiting factor compared to other agroforestry systems (Thevathasan et al., 2018). Accordingly, riparian buffer systems have a greater potential for biomass carbon sequestration with the selection of right mixture of species. Udawatta and Jose (2012) explained that diverse species mixture and the presence of different functional groups such as trees and grasses in riparian buffers and hedgerows enhance the effective capturing of resources which promote carbon sequestration in these systems. Ma et al. (2020) found agroforestry systems with multiple tree species contained greater biomass carbon stocks and accumulated biomass carbon faster than systems with single tree species. This implies that using diverse and complementary species in the implementation of agroforestry systems can yield greater carbon sequestration benefits than using single species.

## 2.2 Belowground (root) biomass carbon sequestration

Roots in agroforestry systems play significant role in long-term carbon sequestration by storing carbon in deeper soil layers (Nair, 2011). For example, roots can constitute up to 50% of total tree and shrub biomass (Axe et al., 2017). Similar to aboveground standing biomass carbon sequestration studies, quantification of belowground biomass carbon in roots of temperate agroforestry systems are scarce. In a meta-analysis to assess the contribution of hedgerows in carbon sequestration, Drexler et al. (2021) observed that only one of the nine studies measured carbon stocks in roots. They suggested the need for more field-based measurements of belowground biomass carbon stocks of hedgerows to derive more thorough estimates. This is because roots contribute substantially to the overall carbon stock of hedgerows.

The observation by Drexler et al. (2021) is not just unique in hedgerows but applies to the other temperate agroforestry systems. Until recently, most root biomass studies in temperate agroforestry focused on competitive interactions between trees and crops for soil resources (e.g. Clinch et al., 2009; Borden et al., 2017). The absence of studies incorporating the potential of root carbon sequestration is due to the major challenges associated with studying root systems. Dube et al. (2012) emphasized that root biomass carbon studies are dominated by logistical challenges including a lack of suitable and appropriate field equipment for excavation and methodological difficulties. For example, during root excavation process, roots break easily, or the soil may not be sufficiently excavated to expose the entire root system (Sinacore et al., 2017). Hence, it requires an enormous effort and time to obtain accurate estimates of root biomass carbon in agroforestry systems (Sinacore et al., 2017). These challenges and efforts required in root biomass studies can serve as disincentives to include root in biomass carbon studies. Alternatively, below-to aboveground conversion factors are usually used to estimate belowground biomass carbon in roots of agroforestry systems (Nair, 2011).

The use of generalized root (belowground) to shoot (aboveground) ratios for belowground carbon estimations is, however, problematic because root to shoot ratios vary widely and can range from 0.20 to 4.50 in temperate biomes, depending on factors such as environmental and edaphic factors, management system, species, stand age, and tree/shrub density (Mokany et al., 2006). Root to shoot ratios of up to 0.50 have been used to estimate belowground biomass carbon in temperate agroforestry systems. For instance, Kort and Turnock (1998) used root to shoot ratios of 0.30, 0.40, and 0.50 to estimate belowground biomass carbon for coniferous, deciduous, and shrub species of shelterbelts in the Canadian prairies. Agroforestry systems with regular aboveground disturbance such as pruning can increase root biomass and root-to-shoot ratio (Drexler et al., 2021). Axe et al. (2017) measured root-to-shoot ratio of 0.94 in hedgerow in the UK. The greater root-to-shoot ratio in hedgerows was attributed to periodic trimming and coppicing of aboveground components (Axe et al., 2017; Drexler et al., 2021). For this reason, using generalized root-to-shoot ratios rather than field quantification of root biomass can result in underestimation of belowground biomass carbon sequestration in temperate agroforestry system, especially where management systems, species and stand densities differ.

Annual root biomass carbon sequestration rates for the different agroforestry systems across temperate regions we have evaluated in our review range from 0.23 Mg C ha$^{-1}$ year$^{-1}$ in alley cropping/intercropping systems to 0.75 Mg C ha$^{-1}$ year$^{-1}$ in hedgerows (Table 2). These root biomass carbon sequestration rates are within the 0.11 to 1.0 Mg C ha$^{-1}$ year$^{-1}$ determined by Cardinael et al. (2018) for temperate agroforestry systems. Our 0.23 Mg C ha$^{-1}$ year$^{-1}$ root carbon rate for alley cropping/intercropping is within the higher range of the 0.11 to 0.23 Mg C ha$^{-1}$ year$^{-1}$ root carbon accumulation rate determined by Wotherspoon et al. (2014) in 25-year-olds tree species in tree-based intercropping systems in southern Ontario. In the riparian buffer system, we calculate 0.57 Mg C ha$^{-1}$ year$^{-1}$ root biomass carbon sequestration rate which is higher than the 0.18 Mg C ha$^{-1}$ year$^{-1}$ observed by Udawatta et al. (2022) for North American riparian buffer systems. Udawatta and Jose (2012) also estimated belowground biomass carbon sequestration rate of 0.09 Mg C ha$^{-1}$ year$^{-1}$ for riparian buffer systems with 50-year harvest cycle.

Root biomass carbon accumulation rates, like in aboveground biomass, depend on the system's age, species, tree density, environmental and edaphic factors, and management impact such as aboveground disturbance (Vijayakumar et al., 2020; Drexler et al., 2021; Viaud and Kunnemann, 2021). Whereas root biomass carbon accumulation can increase with the system's age, the rate of carbon accumulation follows the aforementioned sigmoidal growth pattern of trees (Weiner and Thomas, 2001). Viaud and Kunnemann (2021) observed, in a study in Gourin, northwestern France, that a hedgerow accumulated 4.35 Mg C ha$^{-1}$ in its root biomass after 60 years compared to 3.3 Mg C ha$^{-1}$ and 0.7 Mg C ha$^{-1}$ in a 45- and 20-year-old hedgerow, respectively. Also, the 60-year-old hedgerow, which was pruned 30 years prior to the study, with *Castanea sativa* species was 135% lower in root biomass carbon than the unpruned hedgerow with *Castanea sativa* and *Corylus avellana* in the same location (Viaud and Kunnemann, 2021).

Agroforestry systems such as silvopasture and riparian buffers also show a greater carbon sequestration rate in root biomass than forest and tree plantations. Hazlett et al. (2005) observed 1.84 Mg C ha$^{-1}$ more carbon in belowground biomass of riparian buffers than adjacent boreal forest in Canada. Sharrow and Ismail (2004) also found 1.53 Mg C ha$^{-1}$ in the belowground biomass of 11-year-old Douglas-fir silvopastural system compared to 0.85 Mg C ha$^{-1}$ in the Douglas-fir plantation. Whereas some studies provided evidence of higher carbon sequestration in root biomass of agroforestry systems, others either observed no difference or lower carbon sequestration rate in root biomass of agroforestry systems. For example, López-Díaz et al. (2017) did not observe any significant difference in root biomass carbon in a grazed hybrid walnut (*Juglans major* × *regia* mj 209xra) system in Spain compared to mowed and ploughed systems. Dube et al. (2012), on the other hand, found 82% more carbon in root biomass in a ponderosa pine (*Pinus ponderosa*) plantation compared to a pine-based silvopasture system in Patagonia, Chile. However, tree density was two times greater in the pine plantation compared to the silvopasture system (Dube et al., 2012). This suggests that difference in tree density between the two systems may have played an important role with the lower carbon sequestration rate in the silvopasture system than the pine plantation. Among the factors that impacted carbon sequestration potential of agroforestry systems, Nair and Nair (2003) concluded that tree density played a significant role. It is, therefore, important that the design and implementation of any agroforestry system takes into consideration the optimum tree density required to yield maximum carbon sequestration benefits.

## 2.3 Soil organic carbon sequestration

Total carbon in the terrestrial ecosystem is estimated at 2860 Pg (Lal, 2008). Of this amount ~80% (2300 Pg) is contained in the soil to 1 m depth, with SOC constituting 1550 Pg C (Lal, 2008). Batjes (2014) estimates SOC to 0.3 m depth at 700 Pg C. The 2300 Pg total carbon contained in the soil to a 1 m depth is three times greater than that in the atmosphere (760 Pg C) and 4.1 times greater than that stored in vegetation (560 Pg C) (Lal, 2008). Therefore, the soil represents an important component for carbon storage in terrestrial ecosystems.

Stabilizing atmospheric $CO_2$ will require land management systems that reduce soil carbon loss and enhance SOC storage (Lorenz and Lal, 2014). Among these land management strategies are agroforestry systems which have been recognized under afforestation and reforestation programs for soil carbon sequestration (Nair et al., 2009). Compared to treeless systems, Ma et al. (2020) determined that agroforestry systems increased SOC stocks by an average of 26%. Higher soil carbon sequestration rates in the soil of agroforestry systems was attributed to an increase in organic matter input from above- and belowground components to the soil through litterfall, root turnover and rhizodeposition (Oelbermann et al., 2004). For example, Cardinael et al. (2012) determined that SOC increased by 48% in a willow alley cropping system in southern Canada compared to a 27% increase in the willow monocrop.

Soil carbon storage in agroforestry systems can range from 30 to 300 Mg C ha$^{-1}$ to 1 m depth and depends on the type of agroforestry system, soil properties, and the quantity and quality of organic input into the soil from within the system (Nair et al., 2010). Ofosu et al. (2022) even reported 466.3 Mg C ha$^{-1}$ in tree buffer systems in southern Canada to a 60 cm soil depth compared to 132.2 Mg C ha$^{-1}$ in an adjacent agricultural field. A meta-analysis by Mayer et al. (2022) in temperate agroforestry systems revealed that SOC stocks were 47 Mg C ha$^{-1}$ (0–20 cm) and 32 Mg C ha$^{-1}$ (20–40 cm) compared to 40 Mg C ha$^{-1}$ (0–20 cm) and 29 Mg C ha$^{-1}$ (20–40) in adjacent treeless systems. Soil organic carbon sequestration rates of 1–10 Mg C ha$^{-1}$ year$^{-1}$ (0–50 cm) were also reported by Crossland (2015) for unmanaged hedgerows in England.

Generally, soil carbon sequestration rates in agroforestry systems are higher in upper soil horizons than in deeper soil horizons (Cardinael et al., 2015; Mayer et al., 2022). For example, Mayer et al. (2022) found a SOC sequestration rate of 0.21 Mg C ha$^{-1}$ year$^{-1}$ at a depth of 0–20 cm compared to 0.15 Mg C ha$^{-1}$ year$^{-1}$ at a depth of 20–40 cm in temperate agroforestry systems. Across eight different riparian buffer types and ages in southern Canada, a study by Vijayakumar et al. (2020) revealed SOC stocks in the 0–30 cm depth averaged 135.64 Mg C ha$^{-1}$ compared to 114.68 Mg C ha$^{-1}$ in the 30–60 cm depth. However, the amount of carbon a soil stores depends on the previous land use management practices and the type of agroforestry system that is established (Mayer et al., 2022). Analyzing soil carbon sequestration rates in three different temperate agroforestry systems, Mayer et al. (2022) observed that hedgerows had the highest SOC sequestration rates in 0–20 cm (0.32 Mg C ha$^{-1}$ year$^{-1}$) and 20–40 cm depths (0.28 Mg C ha$^{-1}$ year$^{-1}$), followed by alley cropping systems with 0.26 Mg C ha$^{-1}$ year$^{-1}$ (0–20 cm) and 0.23 Mg C ha$^{-1}$ year$^{-1}$ (20–40 cm) whereas a SOC loss occurred in silvopasture systems at 0–20 cm (–0.17 Mg C ha$^{-1}$ year$^{-1}$) and 20–40 cm (–0.03 Mg C ha$^{-1}$ year$^{-1}$). Ma et al. (2020) observed that converting agricultural land to agroforestry systems yielded a greater soil carbon benefit than converting grassland to agroforestry. As such, it was possible that the SOC loss observed in the silvopasture system by Mayer et al. (2022) was because the adjacent land use for the silvopasture was grassland. For example, Sharrow and Ismail (2004) found a decrease in SOC stocks to a 45 cm depth when pasture (103 Mg C ha$^{-1}$) was converted to silvopasture (96 Mg C ha$^{-1}$) in Oregon, USA. Cardinael et al. (2018) also observed SOC losses ranging from –0.60 to –0.34 Mg C ha$^{-1}$ year$^{-1}$ when grasslands were converted to silvopasture systems in temperate regions across the world. Hence, it is important that comprehensive information about previous land use and land use history is provided in SOC studies in agroforestry systems. This will provide greater and better understanding on how converting different land use to agroforestry systems can impact long-term soil carbon benefits.

Nair (2011) also explained that soil carbon storage in agroforestry systems changes with distance from the tree. For example, Cardinael et al. (2015) observed higher carbon addition to 0–50 cm depth within the tree row (near the trees) compared to 0–30 cm depth in the tree inter-row (away from trees). Higher SOC within tree rows was linked to the lack of soil disturbance in these zones, which created conditions similar to undisturbed systems (Cardinael et al., 2015). Thevathasan et al. (2018), however, explained that the spatial distribution of SOC in alley crops is age related. They suggested that as trees grow, more litter is distributed further from the tree (Thevathasan et al., 2018). In a poplar tree-based intercropping system in southern Canada, Thevathasan and Gordon (1997) determined that 80% of litter fell within 2.5 m of the tree rows with 20% falling in the rest of the field when poplar trees were 7–9 years old. Conversely, only 47% of the litter fell within 2.5 m of the trees when the poplar trees matured to 15 years (Thevathasan et al., 2018). Oelbermann et al. (2006), on the other hand, found higher carbon inputs from litterfall at 1 m (9.46 Mg C ha$^{-1}$) distance from tree row of a 13-year-old hybrid poplar alley cropping system in Ontario, Canada, than at 11.5 m (2.19 Mg C ha$^{-1}$) from tree row. These conflicting results suggest that the processes involved in the spatial distribution of SOC in alley cropping systems are still not well understood. Therefore, long-term field studies are required to fully understand what influences the spatial distribution of SOC in alley cropping systems to inform appropriate soil sampling protocols in agroforestry systems.

Although, soil carbon benefits in agroforestry systems have been widely evaluated, most of these studies took place at a 0–50 cm soil depth (Sharrow and Ismail, 2004; Bambrick et al., 2010; Biffi et al., 2022). This is of concern because roots from woody species extend beyond 50 cm depth and thus, the effect of agroforestry systems on soil carbon stocks may not have been accurately captured (Nair, 2011). Cardinael et al. (2015) observed that out of the 209 Mg C ha$^{-1}$ SOC stocks measured in the 0–200 cm soil depth in an alley cropping system in Montpellier, France, only 40 Mg C ha$^{-1}$, representing less than 20% of the SOC stocks, was contained in the 0–30 cm layer. Haile et al. (2008) also found higher SOC contribution from trees in a slash pine (*Pinus elliottii*) and bahiagrass (*Paspalum notatum*) silvopasture system throughout a 125 cm deep soil profile compared to a non-treed pasture in Florida, USA. The SOC stocks in the silvopasture system were 491 Mg C ha$^{-1}$ (0–50 cm) and 101 Mg C ha$^{-1}$ (50–125 cm) compared to 450 Mg C ha$^{-1}$ (0–50 cm) and 62 Mg C ha$^{-1}$ (50–125 cm) for the non-treed pasture (Haile et al., 2008). However, the impact of tree roots on soil carbon storage in deeper soil layers may not always be positive. Although, Upson and Burgess (2013) determined higher SOC stocks (0–40 cm) in a poplar alley cropping system than in an arable system without trees in England, the reverse was observed for deeper soil layers up to 150 cm. This implies that inadequate sampling depths in agroforestry systems can result in an overestimation of the soil carbon sequestration potential. For instance, from our calculated SOC sequestration rates (Table 2), the few studies that report SOC stocks to depths greater than 50 cm show marginal sequestration rates with even lower SOC rates in lower depths. In riparian buffers, the SOC rate within the upper 50 cm is 3.1 Mg C ha$^{-1}$ year$^{-1}$ compared to 2.8 Mg C ha$^{-1}$ year$^{-1}$ for the upper 100 cm (Table 2). Similarly, silvopasture systems show a higher SOC rate in the upper 50 cm depth (0.5 Mg C ha$^{-1}$ year$^{-1}$) than the upper 100 cm (0.2 Mg C ha$^{-1}$ year$^{-1}$) (Table 2).

Upson and Burgess (2013) explained that at lower depths, a positive priming effect from root exudates could cause a depletion of old SOC. Also, conditions may be more favorable at greater soil depths for enhanced organic matter oxidation (Upson and Burgess, 2013). Whether SOC accumulates or decreases in lower soil depths will depend on several factors including clay content of the soil. For example, Mayer et al. (2022) found that clay content had a positive impact on SOC at lower depths in temperate agroforestry systems. More field studies are required to fully understand the SOC sequestration rate of the various temperate agroforestry systems at greater soil depths. Furthermore, soil has a finite capacity (saturation) to assimilate carbon (Chung et al., 2010), and therefore SOC sequestration in agroforestry systems will not continue indefinitely. Ma et al. (2020) observed that on average, the soil carbon sequestration rate was higher in tropical agroforestry systems than in temperate agroforestry systems. It will generally take 45 years for soils in temperate agroforestry systems to reach equilibrium compared to 14 years in tropical agroforestry systems (Oelbermann, 2002). This implies that temperate agroforestry systems can continue to store more carbon in the soil over the long-term.

Managing agroforestry systems to increase SOC sequestration should consider increasing species diversity and tree density (Nair et al., 2009; Ma et al., 2020). For instance, integrating nitrogen-fixing trees can improve biomass productivity and improve SOC sequestration and stabilization (Kumar et al., 1998; Prescott, 2010). Ma and Chen (2016) determined that enhancing species diversity and richness in forests, grasslands, and croplands increased fine root biomass by 28% and annual production by 45% compared to their monoculture stands. Although, increasing species richness and tree density can increase SOC sequestration (Nair et al., 2009), interspecific competition for growth resources from increased tree density and species can affect biomass production, impacting SOC sequestration (Lorenz and Lal, 2018). The choice of complimentary species and the optimum tree density in the design of temperate agroforestry systems is critical for SOC sequestration and this needs further empirical and field studies.

## 2.4 Soil organic carbon stabilization

Not only do tree-based systems contribute to higher carbon accumulation than treeless systems, but trees also promote the stability of the contributed carbon to the soil by enhancing long-term stabilization of SOM (Haile et al., 2008; Baah-Acheamfour et al., 2014; Vijayakumar et al., 2023). Lorenz and Lal (2014) proposed that to successfully use agroforestry systems as a climate change mitigation strategy, a reduction in SOC loss, and enhanced SOC stabilization is imperative. With SOM being the largest reservoir of SOC in the soil, any change in the size and the turnover rate of SOM impacts SOC pools (Lehmann and Kleber, 2015).

The mechanisms involved in SOM stabilization are not well understood. Different processes and pathways have been proposed to control SOM preservation and stabilization (e.g. Lehmann et al., 2020; Cotrufo and Lavallee, 2022). SOM stabilization is facilitated by biochemical recalcitrance, physical protection, and chemical stabilization through organo-mineral interaction with the soil matrix and environmental factors (Cotrufo and Lavallee, 2022). These stabilization processes limit microbial access to SOM and inhibit the physiological and metabolic capabilities of the soil microbial community involved in plant litter decomposition (Lehmann et al., 2020; Cotrufo and Lavallee, 2022). The biochemical recalcitrance in SOM stabilization is defined by the composition of complex mixtures of organic compounds, which vary in their degradability (Cotrufo and Lavallee, 2022). These complex mixtures of organic compounds may result from the inherent property of the plant litter or are formed during organic matter decomposition (Lorenz and Lal, 2005). Although, the importance of biochemical recalcitrance in SOM stabilization has been questioned (Lehmann and Kleber, 2015), von Lutzow et al. (2006) explained that biochemical recalcitrance is only important during the early stages of decomposition and in active surface soils. Lorenz and Lal (2014) explained that physical protection of SOM and organo-mineral and metal interactions are the two most important groups of processes for long-term SOC stability. This is because physical protection may preserve organic inputs for decades to centuries, whereas organo-mineral complexes or organo-metal interactions may be responsible for stabilization of most of the non-charred SOC for centuries to millennia (Lorenz and Lal, 2014). However, because biochemical recalcitrance plays a crucial role in preserving organic inputs at the early stages of decomposition and in active surface soils (von Lutzow et al., 2006), this delay in organic matter decomposition gives time for physical and organo-mineral interactions to form (Six et al., 2000). As such, mechanisms in SOC stabilization in agroforestry systems involves the combination of both short- and long-term processes (Nair et al., 2010).

Studies evaluating SOM chemistry in temperate agroforestry systems are lacking. Dhillon et al. (2017) reported varying effects of tree species on SOM chemistry in shelterbelts in Saskatchewan, Canada. Whereas some species increased labile carbon forms such as carbohydrates, other species were associated with a higher abundance of recalcitrant aliphatic carbon forms (Dhillon et al., 2017). Compared to annual row cropland, An et al. (2021) and An et al. (2023) observed soil under cropland had more stable SOC than those under temperate agroforestry systems of hedgerows and shelterbelts in western Canada. They explained that SOC may be comparatively higher under agroforestry systems than in cropland (An et al., 2023). Carbon in the agroforestry systems is, however, more susceptible to losses due to higher presence of labile carbon pools (An et al., 2023). Between the hedgerow and shelterbelt, An et al. (2023) observed that the SOC under the hedgerow was more stable than shelterbelt and attributed this to differences in species composition. Tree species diversity can increase SOM degradation by influencing soil microbial community and structure (Lehmann et al., 2020). This implies that SOC stabilization of root-derived carbon through physicochemical interactions in deeper soil horizons may be more important in long-term SOC stabilization under temperate agroforestry systems than the biochemical recalcitrance of aboveground biomass-derived carbon, especially, in systems with high species diversity. Jackson et al. (2017) explained that root inputs are approximately five times more likely to be stabilized as SOC by association with aggregates or clay minerals than an equivalent mass of aboveground litter carbon. Therefore, evaluating the association of soil carbon with different soil size fractions in temperate agroforestry systems will be an important step to understanding their potential impact on long-term SOC stabilization.

Using stable carbon isotope techniques, Haile et al. (2008) observed that slash pine silvopasture systems across four sites in Florida, USA, contributed more carbon to the silt- and clay-sized soil fractions in deeper soil layers compared to an open pasture (no trees). This signified that the trees contributed more stable carbon than the open pasture. Baah-Acheamfour et al. (2014) also studied SOC distribution in three soil fractions (silt- and clay-sized <53 µm, micro-sized 53–250 µm, and macro-sized fractions 250–2000 µm) in temperate agroforestry systems with hedgerows, shelterbelts, and silvopastures across 35 sites in Alberta, Canada. The authors observed a higher SOC content in all the soil fractions in the agroforestry systems (82 g kg$^{-1}$) compared to treeless agricultural systems (54 g kg$^{-1}$). Carbon in the silt- and clay fractions, which constituted the most stable SOC, was 17% higher in the agroforestry systems than the agricultural system (Baah-Acheamfour et al., 2014). Among the agroforestry systems, SOC in the three soil fractions combined was highest in the silvopasture system (82 g kg$^{-1}$), followed by hedgerow (67 g kg$^{-1}$), and lowest in the shelterbelt (50 g kg$^{-1}$). Although shelterbelts had the lowest SOC in all three soil fractions, 58% of its C was located in the silt- and clay-sized fractions (Baah-Acheamfour et al., 2014). This was followed by hedgerow (52%) and the silvopasture (36%). The results from the study by Baah-Acheamfour et al. (2014) imply that soil conditions and litter quality, rather than quantity, may have influenced SOC stabilization in the shelterbelts compared to the silvopasture system.

Also, Oelbermann et al. (2006) observed a greater average annual SOC accumulation rate in a tropical alley cropping system in Costa Rica, due to greater biomass input, compared to a temperate alley cropping system in southern Canada. The temperate system, however, was more efficient in stabilizing the carbon added from crop residues and leaf litter than the tropical system (Oelbermann et al., 2006). This was because SOM mineralization was slower in the cold temperate system compared to the warm and humid tropical system (Oelbermann et al., 2006). Cotrufo and Lavallee (2022) explained that cold conditions inhibit the physiological functions of the soil microbial community, enhancing SOC stabilization. Therefore, climatic conditions in temperate regions could make a substantial contribution to enhancing SOC stability compared to tropical agroforestry systems. More field studies are required to fully understand the mechanisms involved in SOC stabilization under temperate agroforestry systems and their implications for long-term SOC sequestration.

# 3 Projections of potential carbon sequestration in temperate agroforestry systems

Currently, there is a high degree of uncertainty regarding the land area available for agroforestry or those currently under agroforestry practices. This uncertainty poses a challenge in estimating precise carbon sequestration potentials for various agroforestry systems on a global scale. Adding to this layer of challenge is the lack of coordinated research and, in some cases, inconsistencies in reporting results from carbon studies in agroforestry systems. Despite these challenges and shortcomings, we have endeavoured to provide projections on potential carbon sequestration in temperate agroforestry types globally, drawing from estimates calculated in different studies conducted in temperate regions (Table 1). It is crucial to emphasize that these are projections and should be regarded as such.

As in many regions, data on land under agroforestry management in Canada is lacking. Drever et al. (2021) attempted and estimated potential land area available for some land management options for climate change mitigation purposes in Canada, using available spatial data. Among these land options were temperate agroforestry systems of riparian buffer, silvopasture, and alley cropping. The authors projected that across the provinces of Ontario and Quebec (which together occupy approximately 25% of Canada's total land area, as per Statistics Canada (2016)), a total of 797 298 ha is available for alley cropping. This represents only 2% of the 37.9 M ha of cropped land in Canada in 2021 (Statistics Canada, 2022). If 10% of the 2021 cropped land was used for alley cropping, it could potentially mitigate 0.027 Pg $CO_2$ year$^{-1}$ based on our annual carbon sequestration rate of 1.95 Mg C ha$^{-1}$ year$^{-1}$ to a 100 cm soil depth (Table 2).

Drever et al. (2021) also estimated the availability of 985 518 ha and 465 819 ha for silvopasture systems and riparian buffers, respectively, in Canada. These figures do not include areas already under silvopasture and riparian buffer systems. For instance, An et al. (2022) estimated that about 1.8 M ha in the Central Alberta region is already under silvopasture forests. According to the 2021 Canada land use statistics, a total of 4.8 M ha was dedicated to pasture production in Canada (Statistics Canada, 2022). Therefore, if this 4.8 M ha was converted to silvopasture systems, it could potentially help mitigate about 0.026 Pg $CO_2$ year$^{-1}$ using our annual carbon sequestration rate for silvopasture systems of 1.47 Mg C ha$^{-1}$ year$^{-1}$ to a 100 cm soil depth (Table 2). The estimated 465 819 ha available for riparian buffers (Drever et al., 2021) could offset another 0.009 Pg $CO_2$ year$^{-1}$ to a 100 cm soil depth. Moreover, a study by Thevathasan et al. (2018) revealed that approximately 212 000 km of land area in Quebec, Atlantic, and Prairie provinces of Canada were under shelterbelts. An average annual carbon accumulation rate of 3.1 Mg C km$^{-1}$ was calculated by Amichev et al. (2016) at the system-level in prairie shelterbelts. Using this annual rate, the estimated 212 000 km shelterbelt system could potentially mitigate ~0.0024 Pg $CO_2$ annually. Additionally, the 24 390 ha of hedgerow forests in the Central Alberta region alone (An et al., 2022) could contribute a further 0.0008 Pg $CO_2$ year$^{-1}$ in mitigation benefits. Collectively, these examples of temperate agroforestry systems in Canada, while representing only a small fraction of the actual land under agroforestry, constitute approximately 12% (0.065 Pg $CO_2$ year$^{-1}$) of the total 0.546 Pg $CO_2$ emissions for 2021 from all sources in Canada (Global Carbon Project, 2022).

Data from the European agricultural census for 2020 indicate that of the 157 M ha utilized agricultural area in Europe, fodder production occurred on 66 M ha, with the remaining 91 M ha under cereal, kitchen gardens, and industrial crop production (Eurostat, 2020). Also, data from an AGFORWARD project indicate that hedgerows cover about 1.78 M ha in Europe (Burges et al., 2018). If the 66.4 M ha fodder production went under silvopasture system and 10% of the 91 M ha went into silvorable agroforestry such as alley cropping or tree-based intercropping system, using our carbon sequestration rates to 100 cm depth (Table 2) will yield 0.13 Pg C year$^{-1}$ or 0.48 Pg $CO_2$ year$^{-1}$. This yearly $CO_2$ sequestration rate represents ~18% of total $CO_2$ emissions in Europe from all sources for the year 2021.

In the UK, the area under hedgerow is estimated at 158 600 ha (Forestry Commission, 2017). Also, available land for grazing with potential for silvopasture system is 12.4 M ha and about 4.6 M ha land went into arable crop production in 2021 (Government of UK, 2022). If 10% of the 4.6 M ha cropped land went into silvorable system, we estimate 0.02 Pg C year$^{-1}$ or 0.07 Pg $CO_2$ year$^{-1}$, representing ~20% of total $CO_2$ (0.35 Pg $CO_2$) emissions in the UK from all sources for the year 2021 (Global Carbon Project, 2022).

According to estimates by Udawatta et al. (2022), there are potentially 120 M ha of cropland and erodible non-federal land available and suitable for alley cropping in the USA. If 10% of this land (12 M ha) went into alley cropping, using our carbon sequestration rate of 1.95 Mg C ha$^{-1}$ year$^{-1}$ (Table 2) will contribute 0.09 Pg $CO_2$ year$^{-1}$ mitigation benefit. Additionally, the potential 847 500 ha, 34 M ha, and 7.45 M ha available in USA for riparian buffers, silvopastures, and windbreaks, respectively (Udawatta et al., 2022), could potentially contribute a further 0.32 Pg $CO_2$ year$^{-1}$ based on our annual carbon sequestration rates to a 100 cm soil depth (Table 2) for these agroforestry practices. Altogether, these four agroforestry systems could mitigate 0.41 Pg $CO_2$ annually. This value represents ~8.1% of total US $CO_2$ (5.007 Pg $CO_2$) emissions in 2021 (Global Carbon Project, 2022).

Also, 22 M ha of land in the Indian Himalayas has been indicated to be degraded (Kumar et al., 2018). This land could be made productive with the adoption of silvorable systems such as tree-based intercrop. Adoption of tree-based intercrop can yield ~0.16 Pg $CO_2$ year$^{-1}$. The $CO_2$ emissions in India in 2021 is estimated at 2.71 Pg $CO_2$ (Global Carbon Project, 2022). These carbon sequestration rates are indicative of the potential agroforestry practices can have in mitigating climate change through $CO_2$ assimilation.

Due to the slower rate of SOC turnover in colder temperate environments, compared to tropical regions, the benefits of SOC sequestration in temperate agroforestry systems may take a longer period to manifest (Oelbermann and Voroney, 2007). This implies that long-term studies are necessary to understand the carbon sequestration benefits of temperate agroforestry systems. However, conducting long-term research on a decadal or century time scale to evaluate changes in SOC in forest ecosystems and agroecosystems, including agroforestry systems, is impractical. Furthermore, there is a lack of comprehensive long-term and well-designed field studies for the various temperate agroforestry systems to conduct such research. As a result, process-based models have often been used to evaluate and project long-term changes in SOC dynamics under different land management systems. Despite this, there is a limited application of process-based models for biomass and soil carbon studies in temperate agroforestry systems. For example, Oelbermann and Voroney (2011) for the first time calibrated and used the century model to accurately predict long-term SOC changes in a temperate alley cropping system in southern Canada. They found that SOC stocks increased steadily over 100 years once alley cropping systems were initiated, and that most of this carbon accumulated in the slow fraction (Oelbermann and Voroney, 2011). Francaviglia et al. (2012) used the RothC model to evaluate changes in SOC stocks in an agrosilvopastoral system in Italy. They observed that agroforestry systems were more efficient in SOC storage than agricultural systems over a 90-year period (Francaviglia et al., 2012). Additionally, they found that the RothC model accurately predicted SOC changes compared to measured values (Francaviglia et al., 2012). Palma et al. (2018) also demonstrated that modifying the century and RothC models could accurately quantify and predict long-term changes in soil and biomass carbon stocks in hedgerows in the UK. Similarly, Crossland et al. (2015) integrated the RothC into the Yield-SAFE model and accurately quantified soil and biomass carbon stocks in temperate agroforestry systems in the UK and Europe. In both studies by Palma et al. (2018) and Crossland et al. (2015), they found that the biomass and soil carbon values predicted by the models were consistent with measured data. These successful applications of process-based models suggest that they can be used as tools to accurately predict the long-term changes in system-level carbon stocks in temperate agroforestry systems. Thus, it is essential to integrate field studies on system-level carbon stocks with process-based models to determine long-term carbon changes and to further understand potential impacts of agroforestry system management practices on carbon dynamics.

While these examples of temperate agroforestry systems have shown evidence of greater potential for carbon sequestration, significant obstacles persist in their adoption and maintenance on agricultural and other landscapes. A recent study by Amichev et al. (2020) observed that between 2008 and 2016, a total of 2491.2 km of shelterbelts in Saskatchewan, Canada, containing about 190.7 Gt C, had been removed. Farmers have cited the difficulty of operating large farming equipment in crop fields, reduced land area for crop production, and the time, labor, and cost involved in maintaining shelterbelts as some of the major reasons for removing existing shelterbelts in the Canadian prairies (Rempel et al., 2017; Amichev et al., 2020; Drever et al., 2021). A 2014 shelterbelt survey in Manitoba and Saskatchewan, Canada, found that crop producers were more likely than livestock producers to remove shelterbelts to make room for irrigation pivots and large machinery or to expand fields (Rural Development Institute, 2014). Thevathassan et al. (2018) also highlighted the lack of policy incentives and land ownership as additional barriers to adopting and maintaining trees on agricultural lands in Canada. The study by Thevathasan et al. (2018) explained that corporate farms may be less interested in tree planting or other environmental practices that are not perceived as profit-generating activities. Additionally, farming on rented land may pose obstacles to long-term conservation practices involving trees (Thevathasan et al., 2018). The implications of these findings underscore the need for concerted efforts to boost interest in adopting and maintaining agroforestry systems in temperate regions if the full potential benefits of carbon sequestration in these land uses are to be realized.

A proposed solution to increase interest in adopting and maintaining trees on landscapes include the provision of incentives such as financial payments and tax credits to landowners or managers (Thevathasan et al., 2018; Amichev et al., 2020). For example, Mayrinck et al. (2019) argue that implementing carbon market schemes, where any carbon sequestered by agroforestry trees holds a monetary value, would integrate trees into a farm's overall budget. This integration could motivate future tree planting and enhance the maintenance of existing ones. Moreover, it is suggested that farmers may not be fully aware of the advantages of having trees on agricultural landscapes (Rempel et al., 2017). Therefore, education on the long-term environmental benefits of agroforestry practices could help farmers to better value trees on farmlands, leading to increased adoption of agroforestry as a land management option (Mayrinck et al., 2019). A study by Reimer and Prokopy (2014) also revealed that, in addition to financial and environmental benefits, the provision of expertise that reduced technical barriers was equally instrumental to farmers' participation in the USA Farm Bill conservation programs. Thevathasan et al. (2018) suggest the training of high-qualified personnel to provide technical expertise will be required to promote the adoption of agroforestry practices on farms in Canada.

# 4 Conclusion and future directions

Temperate agroforestry practices have the potential to provide climate change mitigation benefits through carbon sequestration, making them a promising option for agricultural landscapes. However, the carbon sequestration benefits of agroforestry systems depend on various factors such as woody species, density, age, management practices, and local environmental and edaphic factors. Analysis of global data from different temperate agroforestry systems reveals that hedgerows have the highest annual carbon sequestration rate (9.35 Mg C ha$^{-1}$ year$^{-1}$) at the system level to 100 cm depth, followed by riparian buffers (5.15 Mg C ha$^{-1}$ year$^{-1}$), shelterbelts/windbreaks (4.48 Mg C ha$^{-1}$ year$^{-1}$), alley cropping/intercropping (1.95 Mg C ha$^{-1}$ year$^{-1}$), and silvopasture systems (1.47 Mg C ha$^{-1}$ year$^{-1}$). Biomass (above- and belowground) accounts for more than half of the system-level carbon sequestration rate in the various agroforestry systems, indicating the need for more attention to biomass carbon quantification. Based on our calculated carbon sequestration rates from various studies conducted in different agroforestry systems for temperate regions, we project that agroforestry practices could annually offset approximately 20%, 18%, 12%, and 8% of total $CO_2$ emissions in the UK, Europe, Canada, and the USA, respectively. We emphasize that the number of reviewed literatures does not provide sufficient information to accurately estimate reliable values for carbon sequestration rates. Additionally, there is a huge degree of uncertainty in the available land area for agroforestry practices. Hence, these projections are just projections and should be regarded as such. Also, agroforestry can contribute to stable carbon pools in the soil, more field studies are required to understand how different agroforestry systems can affect long-term SOC stabilization and the underlying mechanisms. Furthermore, integration of data from field studies into process-based models will help understand how various types of agroforestry systems impact long-term system-level carbon dynamics.

Even though there is increasing research on the carbon sequestration benefits of temperate agroforestry systems, more work is required. The following are some areas that need further consideration to improve our understanding of the carbon sequestration benefits of temperate agroforestry systems.

1 In most studies, crucial information such as tree age, species, tree density, and tree spacing are not provided or clearly defined. Moreover, land management history before the conversion to agroforestry is sometimes not provided. In the absence of baseline data, the inclusion of adjacent land use systems can be useful in estimating and reporting carbon sequestration rates rather than stocks of a particular agroforestry system. This can provide better understanding of a system's potential in mitigating climate change.

2 The importance of temperate agroforestry systems in carbon sequestration has gained greater prominence in the Himalayas in India and North America than other temperate regions. Therefore, more carbon sequestration studies in temperate regions, especially in Europe should be given equal attention.

3 Among the various recognized agroforestry systems in temperate regions, data on forest farming is limited. This may arise from the difficulty in delineating what constitute forest farming from other agroforestry practices. For instance, harvesting sap from maple trees in riparian buffers for maple syrup production is a common practice in southern Ontario, Canada. Managing maple trees for syrup can be classified as forest farming whereas their occupation in riparian zones make them riparian buffers. In short, forest farming has not received the studies other systems have received and more efforts are needed in understanding carbon sequestration benefits of these systems.

4 There is currently no reliable data on how much land area is under agroforestry systems or available for agroforestry practices in temperate regions. This makes it challenging in providing precise estimates on the carbon sequestration potential of these systems. Reliable land area estimates are therefore required to make it possible in estimating the potential carbon sequestration benefits of temperate agroforestry systems.

5 Most agroforestry systems do not account for carbon stocks in companion crops, litterfall, and standing biomass. These components contribute to carbon accumulation at the system-level. Udawatta et al. (2022) equally mention the lack of greenhouse gas emission and source-sink data from agroforestry systems. These data are useful in determining the carbon dynamics and net gains for the adoption and implementation of an agroforestry systems. Hence, more field studies are needed to quantify greenhouse gas emissions in temperate agroforestry systems.

6 The only consistency in soil carbon quantification in agroforestry systems is the inconsistency in soil sampling depths and sometimes reporting soil carbon results. Whereas some studies report in carbon stocks (e.g. Mg ha$^{-1}$), others report carbon content (g kg$^{-1}$ or % C) without providing important parameters such as soil bulk density or soil weight. This makes it difficult to compare results. Even if reporting in carbon concentrations, soil depth and bulk density values or soil weight must be specified to allow for easy comparison with other studies.

## 5   Where to look for further information

The following books and articles provide a good overview of temperate agroforestry systems and carbon benefits:

- Nair, P. R., Kumar, B. M. and Nair, V. D. (2021), 'An introduction to agroforestry: four decades of scientific developments'. 2nd Edition, Springer Nature, pp. 1–666.
- Garrett, H. E. G., Jose, S. and Gold, M. A. (2022), 'North American Agroforestry'. Vol. 194. John Wiley and Sons, pp. 1–566.
- Gordon, A.M., Newman, S.M. and Coleman, B.R.W. (2018), 'Temperate Agroforestry Systems'. 2nd Edition, CABI International, Wallingford, UK, pp. 1-313.
- Lorenz, K. and Lal, R. (2014), 'Soil organic carbon sequestration in agroforestry systems. A review', *Agronomy for Sustainable Development*, 34, 443–454.
- Nair, P. K. R and Garrity, D. (2012), 'Agroforestry - The future of global land use', *Advances in Agroforestry 9*. Springer Netherlands, pp. 1–541.
- Nair, P. K. R. and Nair, V. D. (2003), 'Carbon storage in North American agroforestry systems'. In J. Kimble, L. S. Heath, R. A. Birdsey, and R. Lal (Eds), *The Potential of U.S. Forest Soils to Sequester Carbon and Mitigate the Greenhouse Effect*, CRC Press, Boca Raton, USA, pp. 333–346.
- Udawatta, R.P. and Jose, S. (2012), 'Agroforestry strategies to sequester carbon in temperate North America', *Agroforestry Systems*, 86, 225–42.

Key research and up-to-date research on soil carbon benefits of temperate agroforestry systems can be found at the following organizations:

- Association for Temperate Agroforestry (https://www.aftaweb.org/)
- European Agroforestry Federation (http://www.europeanagroforestry.eu/)

## 6   References

Amichev, B. Y., Bentham, M. J., Kurz, W. A., Laroque, C. P., Kulshreshtha, S., Piwowar, J. M. and van Rees, K. C. (2016), 'Carbon sequestration by white spruce shelterbelts in Saskatchewan, Canada: 3PG and CBM-CFS3 model simulations', *Ecological Modelling*, 325, 35–46.

Amichev, B. Y., Laroque, C. P. and van Rees, K. C. (2020), 'Shelterbelt removals in Saskatchewan, Canada: implications for long-term carbon sequestration', *Agroforestry Systems*, 94, 1665–1680.

Amundson, R. (2022), 'Soil biogeochemistry and the global agricultural footprint', *Soil Security*, 6, 100022. https://doi.org/10.1016/j.soisec.2021.100022

An, Z., Bernard, G. M., Ma, Z., et al. (2021), 'Forest land-use increases soil organic carbon quality but not its structural or thermal stability in a hedgerow system', *Agriculture, Ecosystems & Environment*, 321, 107617.

An, Z., Bork, E. W., Duan, X., Gross, C. D., Carlyle, C. N. and Chang, S. X. (2022), 'Quantifying past, current, and future forest carbon stocks within agroforestry systems in central Alberta, Canada', *GCB Bioenergy*, 14(6), 669–680.

An, Z., Pokharel, P., Plante, A. F., Bork, E. W., Carlyle, C. N., Williams, E. K. and Chang, S. X. (2023), 'Soil organic matter stability in forest and cropland components of two agroforestry systems in western Canada', *Geoderma*, 433, 116463.

Axe, M. S., Grange, I. D. and Conway, J. S. (2017), 'Carbon storage in hedge biomass – A case study of actively managed hedges in England', *Agriculture, Ecosystems & Environment*, 250, 81–88.

Baah-Acheamfour, M., Carlyle, C. N., Bork, E. W., Chang, S. X. (2014), 'Trees increase soil carbon and its stability in three agroforestry systems in central Alberta, Canada', *Ecological Management*, 328, 131–139.

Bainbridge, D. A., Virginia, R. A. and Jarrell, W. M. (1990), 'Honey mesquite: a multi-purpose tree for arid lands', *Nitrogen Fixing Tree Association*, 90(7), 2–4.

Balian, E. V. and Naiman, R. J. (2005), 'Abundance and production of riparian trees in the lowland floodplain of the Queets River, Washington', *Ecosystems*, 8, 841–861.

Bambrick, A. D., Whalen, J. K., Bradley, R. L., Cogliastro, A., Gordon, A. M., Olivier, A. and Thevathasan, N. V. (2010), 'Spatial heterogeneity of soil organic carbon in tree-based intercropping systems in Quebec and Ontario, Canada', *Agroforestry Systems*, 79, 343–353.

Batjes, N. H. (2014), 'Total carbon and nitrogen in the soils of the world', *European Journal of Soil Science*, 65(1), 10–21.

Bhardwaj, D. R., Salve, A., Kumar, J., Kumar, A., Sharma, P. and Kumar, D. (2023), 'Biomass production and carbon storage potential of agroforestry land use systems in high hills of north-western Himalaya: an approach towards natural based climatic solution', *Biomass Conversion and Biorefinery*, 1–14.

Biffi, S., Chapman, P. J., Grayson, R. P. and Ziv, G. (2022), 'Soil carbon sequestration potential of planting hedgerows in agricultural landscapes', *Journal of Environmental Management*, 307, 114484.

Borden, K. A., Isaac, M. E., Thevathasan, N. V., Gordon, A. M. and Thomas, S. C. (2014), 'Estimating coarse root biomass with ground penetrating radar in a tree-based intercropping system', *Agroforestry Systems*, 88, 657–669.

Borden, K. A., Thomas, S. C. and Isaac, M. E. (2017), 'Interspecific variation of tree root architecture in a temperate agroforestry system characterized using ground-penetrating radar', *Plant and Soil*, 410, 323–334.

Burgess, P. J., den Herder, M., Dupraz, C., Garnett, K., Giannitsopoulos, M., Graves, A. R., Hermansen, J. E., Kanzler, M., Liagre, F., Mirck, J., Moreno, G., Mosquera-Losada, M. R., Palma, J. H. N., Pantera, A. and Plieninger, T. (2018), 'AGFORWARD PROJECT Final Report. Cranfield University: AGFORWARD', https://www.agforward.eu/documents/AGFORWARD%20Final%20Report%2028%20Feb%202018.pdf (Accessed April 17, 2023).

Cardinael, R., Chevallier, T., Barthès, B. G., et al. (2015), 'Impact of alley cropping agroforestry on stocks, forms and spatial distribution of soil organic carbon—A case study in a Mediterranean context', *Geoderma*, 259, 288–299.

Cardinael, R., Chevallier, T., Cambou, A., et al. (2017), 'Increased soil organic carbon stocks under agroforestry: a survey of six different sites in France', *Agriculture, Ecosystems & Environment*, 236, 243–255.

Cardinael, R., Thevathasan, N., Gordon, A., Clinch, R., Mohammed, I. and Sidders, D. (2012), 'Growing woody biomass for bioenergy in a tree-based intercropping system in southern Ontario, Canada', *Agroforestry Systems*, 86, 279–286.

Cardinael, R., Umulisa, V., Toudert, A., Olivier, A., Bockel, L. and Bernoux, M. (2018), 'Revisiting IPCC Tier 1 coefficients for soil organic and biomass carbon storage in agroforestry systems', *Environmental Research Letters*, 13(12), 124020.

Castleden, H., Garvin, T. and Huu-ay-aht First Nation (2009), 'Hishuk Tsawak" (Everything is one/connected): a Huu-ay-aht worldview for seeing forestry in British Columbia, Canada', *Society & Natural Resources*, 22(9), 789–804. https://doi.org/10.1080/08941920802098198

Chang, S. X., Wang, W., Zhu, Z., Wu, Y. and Peng, X. (2018), 'Temperate agroforestry in China'. In A. M. Gordon, S. M. Newman ad B. R. W. Coleman (Eds), *Temperate Agroforestry Systems*. 2nd Edition, CAB International, Wallingford, UK, pp. 173–194.

Chisanga, K., Bhardwaj, D. R., Pala, N. A. and Thakur, C. L. (2018), 'Biomass production and carbon stock inventory of high-altitude dry temperate land use systems in North Western Himalaya', *Ecological Processes*, 7(1), 1–13.

Chu, X., Zhan, J., Li, Z., Zhang, F. and Qi, W. (2019), 'Assessment on forest carbon sequestration in the Three-North Shelterbelt Program region, China', *Journal of Cleaner Production*, 215, 382–389.

Chung, H., Ngo, K. J., Plante, A. and Six, J. (2010), 'Evidence for carbon saturation in a highly structured and organic-matter-rich soil', *Soil Science Society of America Journal*, 74(1), 130–138.

Crossland, M. (2015), 'The carbon sequestration potential of hedges managed for woodfuel', *The Organic Research Centre, Elm Farm*. https://www.organicresearchcentre.com/manage/authincludes/article_uploads/project_outputs/TWECOM%20ORC%20Carbon%20report%20v1.0.pdf (Accessed April 17, 2023).

Clinch, R. L., Thevathasan, N. V., Gordon, A. M., Volk, T. A. and Sidders, D. (2009), 'Biophysical interactions in a short rotation willow intercropping system in southern Ontario, Canada', *Agriculture, Ecosystems & Environment*, 131(1–2), 61–69.

Cotrufo, M. F. and Lavallee, J. M. (2022), 'Soil organic matter formation, persistence, and functioning: a synthesis of current understanding to inform its conservation and regeneration', *Advances in Agronomy*, 172, 1–66.

Dhillon, G. S., Gillespie, A., Peak, D. and van Rees, K. C. (2017), 'Spectroscopic investigation of soil organic matter composition for shelterbelt agroforestry systems', *Geoderma*, 298, 1–13.

Dixon, R. K. (1995), Agroforestry systems: sources and sinks of greenhouse gases? Agroforestry Systems, 31, 99–116.

Dixon, R. K., Winjum, J. K. and Schroeder, P. E. (1993), Conservation and sequestration of carbon: the potential of forest and agroforestry management practices. *Global Environmental Change*, 3, 159–173.

Drake, J. B., Knox, R. G., Dubayah, R. O., Clark, D. B., Condit, R., Blair, J. B. and Hofton, M. (2003), 'Above-ground biomass estimation in closed canopy neotropical forests using lidar remote sensing: factors affecting the generality of relationships', *Global Ecology and Biogeography*, 12(2), 147–159.

Drever, C. R., Cook-Patton, S. C., Akhter, F., et al. (2021), 'Natural climate solutions for Canada', *Science Advances*, 7(23), eabd6034.

Drexler, S., Gensior, A. and Don, A. (2021), 'Carbon sequestration in hedgerow biomass and soil in the temperate climate zone', *Regional Environmental Change*, 21(3), 74.

Dube, F., Espinosa, M., Stolpe, N. B., Zagal, E., Thevathasan, N. V. and Gordon, A. M. (2012), 'Productivity and carbon storage in silvopastoral systems with Pinus ponderosa and Trifolium spp., plantations and pasture on an Andisol in Patagonia, Chile', *Agroforestry Systems*, 86, 113–128.

Dube, F., Stolpe, N. B., Zagal, E., Figueroa, C. R., Concha, C., Neira, P., Carrasco, C., Schwenke, J. M., Schwenke, V. and Müller, B. (2018), 'Novel agroforestry systems in temperate Chile'. In A. M. Gordon, S. M. Newman and B. R. W. Coleman (Eds), *Temperate Agroforestry Systems*. 2nd Edition, CABI International, Wallingford, UK, pp. 237–251.

Dupraz, C., Lawson, G.J., Lamersdorf, N., Papanastasis, V.P., Rosati, A. and Ruiz-Mirazo, J. (2018), 'Temperate agroforestry: the European way'. In A. M. Gordon, S. M. Newman, ad B. R. W. Coleman (Eds), *Temperate Agroforestry Systems*. 2nd Edition, CABI International, Wallingford, UK, pp. 98–152.

Eurostat. (2020), 'Agri-environmental indicators'. https://ec.europa.eu/eurostat/statistics -explained/index.php?title=Agri-environmental_indicator_-_cropping_patterns (Accessed April 17, 2023)

Fernández-Núñez, E., Rigueiro-Rodríguez, A. and Mosquera-Losada, M. R. (2010), 'Carbon allocation dynamics one decade after afforestation with *Pinus radiata* D. Don and *Betula alba* L. under two stand densities in NW Spain', *Ecological Engineering*, 36(7), 876–890.

Forestry Commission. (2017), 'Tree cover outside woodland in Great Britain National Forest Inventory', April 2017. https://cdn.forestresearch.gov.uk/2022/02/fr_tree _cover_outside_woodland_in_gb_statistical_report_2017.pdf (Accessed April 17, 2023).

Fortier, J., Truax, B., Gagnon, D. and Lambert, F. (2013), 'Root biomass and soil carbon distribution in hybrid poplar riparian buffers, herbaceous riparian buffers and natural riparian woodlots on farmland', *Springer Plus*, 2(1), 1–19.

Fortier, J., Truax, B., Gagnon, D. and Lambert, F. (2015), 'Biomass carbon, nitrogen and phosphorus stocks in hybrid poplar buffers, herbaceous buffers and natural woodlots in the riparian zone on agricultural land', *Journal of Environmental Management*, 154, 333–345.

Francaviglia, R., Coleman, K., Whitmore, A. P., Doro, L., Urracci, G., Rubino, M. and Ledda, L. (2012), 'Changes in soil organic carbon and climate change–application of the RothC model in agro-silvo-pastoral Mediterranean systems', *Agricultural Systems*, 112, 48–54.

Giese, L. A., Aust, W. M., Kolka, R. K. and Trettin, C. C. (2003), 'Biomass and carbon pools of disturbed riparian forests', *Forest Ecology and Management*, 180(1–3), 493–508.

Global Carbon Project. (2022), 'Carbon dioxide emissions worldwide in 2010 and 2021, by select country (in million metric tons)', Statista. Statista Inc. https://www.statista .com/statistics/270499/co2-emissions-in-selected-countries/ (Accessed April 17, 2023).

Gordon, A. M. and Thevathasan, N. V. (2004), 'How much carbon can be stored in Canadian agroecosystems using a silvopastoral approach?'. In M. R. Mosquera-Losada, J. McAdam, and A. Rigueiro-Rodriguez (Eds), *Silvopastoralism and sustainable land management*. CAB International, Wallingford, UK. pp. 210–218.

Goswami, S., Verma, K. S. and Kaushal, R. (2014), 'Biomass and carbon sequestration in different agroforestry systems of a Western Himalayan watershed', *Biological Agriculture & Horticulture*, 30(2), 88–96.

Government of UK (2022), 'Agriculture in the United Kingdom 2021', Department for Environment, Food and Rural Affairs. UK. https://assets.publishing.service.gov.uk/government/uploads/system/uploads/attachment_data/file/1094493/Agriculture-in-the-UK-27jul22.pdf (Accessed April 17, 2023).

Gross, C. D., Bork, E. W., Carlyle, C. N. and Chang, S. X. (2022), 'Agroforestry perennials reduce nitrous oxide emissions and their live and dead trees increase ecosystem carbon storage', *Global Change Biology*, 28(20), 5956–5972.

Haile, S. G., Nair, P. R. and Nair, V. D. (2008), 'Carbon storage of different soil-size fractions in Florida silvopastoral systems', *Journal of Environmental Quality*, 37(5), 1789–1797.

Hazlett, P. W., Gordon, A. M., Sibley, P. K. and Buttle, J. M. (2005), 'Stand carbon stocks and soil carbon and nitrogen storage for riparian and upland forests of boreal lakes in northeastern Ontario', *Forest Ecology and Management*, 219(1), 56–68.

Hoffman, K.M., Christianson, A.C., Dickson-Hooyle, S., Copes-Gerbitz, K., Nikolakis, W., Diabo, D.A., McLeod, R., Mitchell, H.J., Al Mamun, A., Zahara, A., Mauro, N., Gilchrist, J., Myers Ross, R. and Daniels, L.D. (2022), The right to burn: barriers and opportunities for Indigenous-led fire stewardship in Canada. *FACETS*, 7, 464–481. https://doi.org/10.1139/facets-2021-0062

Houghton, R. A. and Hackler, J. L. (2000), 'Changes in terrestrial carbon storage in the United States. 1: The roles of agriculture and forestry', *Global Ecology and Biogeography*, 9(2), 125–144.

Jackson, R. B., Lajtha, K., Crow, S. E., Hugelius, G., Kramer, M. G. and Piñeiro, G. (2017), 'The ecology of soil carbon: pools, vulnerabilities, and biotic and abiotic controls', *Annual Review of Ecology, Evolution, and Systematics*, 48, 419–445.

Kaonga, M. L. and Bayliss-Smith, T. P. (2009), 'Carbon pools in tree biomass and the soil in improved fallows in eastern Zambia', *Agroforestry Systems*, 76(1), 37–51.

Kemp, P. D., Hawke, M. F. and Knowles, R. L. (2018), 'Temperate agroforestry systems in New Zealand'. In A. M. Gordon, S. M. Newman, ad B. R. W. Coleman (Eds), *Temperate Agroforestry Systems*. 2nd Edition, CAB International, Wallingford, UK, pp. 224–236.

Ketterings, Q. M., Coe, R., van Noordwijk, M. and Palm, C. A. (2001), 'Reducing uncertainty in the use of allometric biomass equations for predicting above-ground tree biomass in mixed secondary forests', *Forest Ecology and Management*, 146(1–3), 199–209.

King, K.F.S. (1987), 'The history of agroforestry'. In H. A. Steppler and P. K. R. Nair (Eds), *Agroforestry: A Decade of Development*. International Centre for Research in Agroforestry (ICRAF), Nairobi, Kenya, pp. 3–13.

Kort, J. and Turnock, R. (1998), 'Carbon reservoir and biomass in Canadian prairie shelterbelts', *Agroforestry Systems*, 44(2–3), 175–186.

Kumar, B. M., Handa, A. K., Dhyani, S. K. and Arunachalam, A. (2018), 'Agroforestry in the Indian Himalayan region: an overview'. In A. M. Gordon, S. M. Newman, ad B. R. W. Coleman (Eds), *Temperate Agroforestry Systems*. 2nd Edition, CAB International, Wallingford, UK, pp. 153–172.

Kumar, B. M., Kumar, S. S. and Fisher, R. F. (1998), 'Intercropping teak with Leucaena increases tree growth and modifies soil characteristics', *Agroforestry Systems*, 42, 81–89.

Kürsten, E. and Burschel, P. (1993), '$CO_2$-mitigation by agroforestry', *Water, Air, & Soil Pollution*, 70, 533–544.

Lal, R. (2008), 'Carbon sequestration', *Philosophical Transactions of the Royal Society B*, 363, 815–830.

Lehmann, J., Hansel, C. M., Kaiser, C., et al. (2020), 'Persistence of soil organic carbon caused by functional complexity', *Nature Geoscience*, 13(8), 529–534.

Lehmann, J. and Kleber, M. (2015), 'The contentious nature of soil organic matter', *Nature*, 528(7580), 60–68.

Lorenz, K. and Lal, R. (2005), 'The depth distribution of soil organic carbon in relation to land use and management and the potential of carbon sequestration in subsoil horizons', *Advances in Agronomy*, 88, 35–66.

Lorenz, K. and Lal, R. (2014), 'Soil organic carbon sequestration in agroforestry systems: a review', *Agronomy for Sustainable Development*, 34, 443–454.

Lorenz, K. and Lal, R. (2018), *Carbon Sequestration in Agricultural Ecosystems*, Springer International Publishing, Cham.

López-Díaz, M. L., Benítez, R. and Moreno, G. (2017), 'How do management techniques affect carbon stock in intensive hardwood plantations?', *Forest Ecology and Management*, 389, 228–239.

Luick, R. (2009), 'Wood pastures in Germany'. In A. Rigueiro-Rodriguez, J. McAdam, and M. Mosquera-Losado (Eds), *Agroforestry in Europe: Current Status and Future Prospects*. Springer Science, Dordrecht, NL, pp. 359–376.

Ma, Z. and Chen, H. Y. (2016), 'Effects of species diversity on fine root productivity in diverse ecosystems: a global meta-analysis', *Global Ecology and Biogeography*, 25(11), 1387–1396.

Ma, Z., Chen, H. Y., Bork, E. W., Carlyle, C. N. and Chang, S. X. (2020), 'Carbon accumulation in agroforestry systems is affected by tree species diversity, age and regional climate: a global meta-analysis', *Global Ecology and Biogeography*, 29(10), 1817–1828.

Marchildon, G. P. (2009), 'The prairie farm rehabilitation administration: climate crisis and federal–provincial relations during the Great Depression', *Canadian Historical Review*, 90(2), 275–301. https://doi.org/10.3138/chr.90.2.275

Mayer, S., Wiesmeier, M., Sakamoto, E., Hübner, R., Cardinael, R., Kühnel, A. and Kögel-Knabner, I. (2022), Soil organic carbon sequestration in temperate agroforestry systems – A meta-analysis. *Agriculture, Ecosystems & Environment*, 323, 107689. https://doi.org/10.1016/j.agee.2021.107689

Mayrinck, R. C., Laroque, C. P., Amichev, B. Y. and van Rees, K. (2019), 'Above-and below-ground carbon sequestration in shelterbelt trees in Canada: a review', *Forests*, 10(10), 922.

Mokany, K., Raison, R. J. and Prokushkin, A. S. (2006), 'Critical analysis of root: shoot ratios in terrestrial biomes', *Global Change Biology*, 12(1), 84–96.

Montagnini, F., Francesconi, W. and Rossi, E. (2011), *Agroforestry as a Tool for Landscape Restoration*. Nova Science Publishers, UK.

Nair, P. R. (2011), 'Agroforestry systems and environmental quality: introduction', *Journal of Environmental Quality*, 40(3), 784–790.

Nair, P. K. R. and Nair, V. D. (2003), 'Carbon storage in North American agroforestry systems'. In J. Kimble, L. S. Heath, R. A. Birdsey, and R. Lal (Eds), *The Potential of U.S. Forest Soils to Sequester Carbon and Mitigate the Greenhouse Effect*, CRC Press, Boca Raton, USA, pp. 333–346.

Nair, R. P. K., Mohan Kumar, B. and Nair, V. D. (2009), 'Agroforestry as a strategy for carbon sequestration', *Journal of Plant Nutrition and Soil Science*, 172, 10–23. https://doi.org/10.1002/jpln.200800030

Nair, P. K. R., Nair, V. D., Kumar, B. M. and Showalter, J. M. (2010), 'Carbon sequestration in agroforestry systems', *Advances in Agronomy*, 108, 237–307.

Nerlich, K., Graeff-Hönninger, S. and Claupein, W. (2013), 'Agroforestry in Europe: a review of the disappearance of traditional systems and development of modern agroforestry practices, with emphasis on experiences in Germany', *Agroforestry Systems*, 87, 475–492. https://doi.org/10.1007/s10457-012-9560-2

Oelbermann, M. (2002), '*Linking carbon inputs to sustainable agriculture in Canadian and Costa Rican agroforestry systems*', Ph.D. Thesis. Department of Land Resource Science, University of Guelph, p. 208.

Oelbermann, M. and Voroney, R. P. (2007), 'Carbon and nitrogen in a temperate agroforestry system: using stable isotopes as a tool to understand soil dynamics', *Ecological Engineering*, 29(4), 342–349.

Oelbermann, M. and Voroney, R. P. (2011), 'An evaluation of the century model to predict soil organic carbon: examples from Costa Rica and Canada', *Agroforestry Systems*, 82, 37–50.

Oelbermann, M., Voroney, R. P. and Gordon, A. M. (2004), 'Carbon sequestration in tropical and temperate agroforestry systems: a review with examples from Costa Rica and Southern Canada', *Agriculture, Ecosystems & Environment*, 104, 359–377.

Oelbermann, M., Voroney, R. P., Kass, D. C. L. and Schlönvoigt, A. M. (2005), 'Above- and below-ground carbon inputs in 19-, 10-, and 4-year old Costa Rican alley cropping systems', *Agriculture, Ecosystems & Environment,* 105, 163–172.

Oelbermann, M., Voroney, R. P., Thevathasan, N. V., Gordon, A. M., Kass, D. C. and Schlönvoigt, A. M. (2006), 'Soil carbon dynamics and residue stabilization in a Costa Rican and southern Canadian alley cropping system', *Agroforestry Systems*, 68(1), 27–36.

Ofosu, E., Bazrgar, A., Coleman, B., Deen, B., Gordon, A., Voroney, P. and Thevathasan, N. (2022), 'Diverse temperate riparian buffer types promote system-level carbon sequestration in Southern Ontario, Canada', *The Forestry Chronicle*, 98(1), 103–118.

Palma, J. H., Crous-Durán, J., Graves, A. R., et al. (2018), 'Integrating belowground carbon dynamics into Yield-SAFE, a parameter sparse agroforestry model', *Agroforestry Systems*, 92, 1047–1057.

Pardon, P., Reubens, B., Mertens, J., Verheyen, K., De Frenne, P., De Smet, G., Van Waes, C. and Reheul, D. (2018), Effects of temperate agroforestry on yield and quality of different arable intercrops, *Agricultural Systems*, 166, 135-151 https://doi.org/10.1016/j.agsy.2018.08.008.

Peichl, M., Thevathasan, N. V., Gordon, A. M., Huss, J. and Abohassan, R. A. (2006), 'Carbon sequestration potentials in temperate tree-based intercropping systems, Southern Ontario, Canada', *Agroforestry Systems*, 66, 243–257.

Peri, P. L., Caballé, G., Hansen, N. E., Bahamonde, H. A., Lencinas, M. V., von Müller, A. R., Ormaechea, S., Gargaglione, V., Soler, R., Sarasola, M., Rusch, V., Borrelli, L., Fernández, M. E., Gyenge, J., Tejera, L. E., Lloyd, C. E. and Martínez Pastur, G. (2018), 'Silvopastoral systems in Patagonia, Argentina'. In A. M. Gordon, S. M. Newman, and B. R. W. Coleman (Eds), *Temperate Agroforestry Systems*. 2nd Edition, CAB International, Wallingford, UK, pp. 252–273.

Prescott, C. E. (2010), 'Litter decomposition: what controls it and how can we alter it to sequester more carbon in forest soils?', *Biogeochemistry*, 101, 133–149.

Rajput, B. S., Bhardwaj, D. R. and Pala, N. A. (2015), 'Carbon dioxide mitigation potential and carbon density of different land use systems along an altitudinal gradient in north-western Himalayas', *Agroforestry Systems*, 89, 525–536.

Rajput, B. S., Bhardwaj, D. R. and Pala, N. A. (2017), 'Factors influencing biomass and carbon storage potential of different land use systems along an elevational gradient in temperate northwestern Himalaya', *Agroforestry Systems*, 91, 479–486.

Reeg, T., Möndel, A., Brix, M. and Konold, W. (2008), Conservation in agricultural landscape– new options through modern agroforestry systems? *Nature Landscape*, 83, 261–266.

Reid, R. and Moore, R. (2018), 'Agroforestry systems in temperate Australia'. In A. M. Gordon, S. M. Newman and B. R. W. Coleman (Eds), *Temperate Agroforestry Systems*. 2nd Edition, CAB International, Wallingford, UK, pp. 195–223.

Reimer, A. P. and Prokopy, L. S. (2014), 'Farmer participation in US Farm Bill conservation programs', *Environmental Management*, 53, 318–332.

Rempel, J. C., Kulshreshtha, S. N., Amichev, B. Y. and Van Rees, K. C. (2017), 'Costs and benefits of shelterbelts: a review of producers' perceptions and mind map analyses for Saskatchewan, Canada'. *Canadian Journal of Soil Science*, 97(3), 341–352.

Rheinhardt, R., Brinson, M., Meyer, G. and Miller, K. (2012), 'Integrating forest biomass and distance from channel to develop an indicator of riparian condition', *Ecological Indicators*, 23, 46–55.

Rural Development Institute. (2014). '2014 shelterbelt survey prairie producers' use of and attitudes towards shelterbelts', Brandon University. https://www.brandonu.ca/rdi/files/2011/02/AGGP-Survey-Report-2014.pdf (Accessed 25 November 2023).

Sánchez, P.A. (2000), Linking climate change research with food security and poverty reduction in the tropics. *Agriculture, Ecosystems & Environment*, 82, 371–383.

Schroeder, P. (1994), 'Carbon storage benefits of agroforestry systems', *Agroforestry Systems*, 27, 89–98.

Sharma, H., Pant, K. S., Bishist, R., Gautam, K. L., Dogra, R., Kumar, M. and Kumar, A. (2023), 'Estimation of biomass and carbon storage potential in agroforestry systems of north western Himalayas, India', *CATENA*, 225, 107009.

Sharrow, S. H. and Ismail, S. (2004), 'Carbon and nitrogen storage in agroforests, tree plantations, and pastures in western Oregon, USA', *Agroforestry Systems*, 60, 123–130.

Sinacore, K., Hall, J. S., Potvin, C., Royo, A. A., Ducey, M. J. and Ashton, M. S. (2017), 'Unearthing the hidden world of roots: Root biomass and architecture differ among species within the same guild', *PLoS One*, 12(10), e0185934.

Six, J. A. E. T., Elliott, E. T. and Paustian, K. (2000), 'Soil macroaggregate turnover and microaggregate formation: a mechanism for C sequestration under no-tillage agriculture', *Soil Biology and Biochemistry*, 32(14), 2099–2103.

Statistics Canada. (2016), '2016 Census of Agriculture', https://www.statcan.gc.ca/en/ca2016 (Accessed April 17, 2023).

Thevathasan, N. V., Coleman, B., Zabek, L., Ward, T. and Gordon, A. M. (2018), 'Agroforestry in Canada and its role in farming systems'. In A. M. Gordon, S. M. Newman, and B. R. W. Coleman (Eds), *Temperate Agroforestry Systems*. 2nd Edition, CAB International, Wallingford, UK, pp. 7–49.

Thevathasan, N. V., Coleman, B., Zabek, L., Ward, T. and Gordon, A. M. (2018), 'Agroforestry in Canada and its role in farming systems'. In A. M. Gordon, S. M. Newman, and B. R. W. Coleman (Eds), *Temperate Agroforestry Systems*. 2nd Edition, CAB International, Wallingford, UK, pp. 7–49.

Thevathasan, N. V. and Gordon, A. M. (1997), 'Poplar leaf biomass distribution and nitrogen dynamics in a poplar-barley intercropped system in Southern Ontario, Canada', *Agroforestry Systems*, 37, 79–90.

Tufekcioglu, A., Raich, J. W., Isenhart, T. M. and Schultz, R. C. (2003), 'Biomass, carbon and nitrogen dynamics of multi-species riparian buffers within an agricultural watershed in Iowa, USA', *Agroforestry Systems*, 57, 187–198.

Udawatta, R. P., Walter, D. and Jose, S. (2022), 'Carbon sequestration by forests and agroforests: a reality check for the United States', *Carbon Footprints*, 1, 8. https://dx .doi.org/10.20517/cf.2022.06

Udawatta, R.P. and Jose, S. (2012), 'Agroforestry strategies to sequester carbon in temperate North America', *Agroforestry Systems*, 86, 225–42.

Upson, M. A. and Burgess, P. J. (2013), 'Soil organic carbon and root distribution in a temperate arable agroforestry system', *Plant and Soil*, 373, 43–58.

Viaud, V. and Kunnemann, T. (2021), 'Additional soil organic carbon stocks in hedgerows in crop-livestock areas of western France', *Agriculture, Ecosystems & Environment*, 305, 107174.

Vijayakumar, S., Bazrgar, A. B., Coleman, B., Gordon, A., Voroney, P. and Thevathasan, N. (2020), 'Carbon stocks in riparian buffer systems at sites differing in soil texture, vegetation type and age compared to adjacent agricultural fields in southern Ontario, Canada', *Agriculture, Ecosystems & Environment*, 304, 107149.

Vijayakumar, S., Bazrgar, A. B., Gordon, A., Voroney, P. and Thevathasan, N. (2023), 'Soil organic carbon stabilization based on physical fractionation method in tree based riparian and adjacent agricultural systems in southern Ontario, Canada'. *Agroforestry Systems*, 1–15. https://doi.org/10.1007/s10457-023-00913-4

von Lützow, M., Kögel-Knabner, I., Ekschmitt, K., Matzner, E., Guggenberger, G., Marschner, B. and Flessa, H. (2006), 'Stabilization of organic matter in temperate soils: mechanisms and their relevance under different soil conditions–a review', *European Journal of Soil Science*, 57(4), 426–445.

Watson, R.T., Noble, I.R., Bolin, B., Ravindranath, N.H., Verardo, D.J. and Dokken, D.J. (2000), *Land use, Land-use Change and Forestry*. Cambridge University Press, Cambridge, England.

Weiner, J. and Thomas, S. C. (2001), 'The nature of tree growth and the "age-related decline in forest productivity"', *Oikos*, 94(2), 374–376.

Williams, P. A., Gordon, A. M., Garret, H. E. and Buck, L. (1997), 'Agroforestry in North America and its role in farming systems'. In A. M. Gordon and S. M. Newman (Eds), *Temperate Agroforestry Systems*. 1st Edition, CAB International, Wallingford, UK, pp. 9–85.

Winans, K. S., Whalen, J. K., Rivest, D., Cogliastro, A. and Bradley, R. L. (2016), 'Carbon sequestration and carbon markets for tree-based intercropping systems in Southern Quebec, Canada', *Atmosphere*, 7(2), 17.

Worms, P. (2021), A history on the forgotten but enduring practice of agroforestry and the role it can play in shaping the future of Europe's sustainable agriculture. *Revolve*, 41. https://revolve.media/agroforestry-the-age-old-future-of-europes -agriculture/

Wotherspoon, A., Thevathasan, N. V., Gordon, A. M. and Voroney, R. P. (2014), 'Carbon sequestration potential of five tree species in a 25-year-old temperate tree-based intercropping system in Southern Ontario, Canada', *Agroforestry systems*, 88, 631–643.

You, X. L. (1991), 'Mixed cropping with trees in ancient China'. In Z. H. Zhu, M. T. Cai, S. J. Wang, and Y. X. Jiang (Eds), *Agroforestry Systems in China*. IDRC Canada and CAF, China, pp. 8–9.

Young, A. (1997), *Agroforestry for Soil Management*. CAB International, Wallingford, UK.

Zahoor, S., Dutt, V., Mughal, A. H., Pala, N. A., Qaisar, K. N. and Khan, P. A. (2021), 'Apple-based agroforestry systems for biomass production and carbon sequestration: implication for food security and climate change contemplates in temperate region of Northern Himalaya, India', *Agroforestry Systems*, 95, 367–382.

www.ingramcontent.com/pod-product-compliance
Lightning Source LLC
Chambersburg PA
CBHW070736220326
41598CB00024BA/3442